George Wilson's Vision of Early Victorian Science and Technology

This volume is a comprehensive study of George Wilson, a leading advocate for evangelical science and for the role of biology in technology – it examines his work to develop a unitary vision of Victorian science and technology by drawing upon religion, transcendental natural history, and Baconian philosophy.

George Wilson was the first Regius Professor of Technology at the University of Edinburgh and the founding Director of the Industrial Museum of Scotland (now the National Museum of Scotland). Throughout his career, he lectured and published on a wide range of topics, including the prospect of life on other planets, the history of science, natural theology, chemistry, and poetry. His works were very popular – he was praised by Charles Dickens and his lectures drew large audiences, particularly women. Wilson sought to educate people about the significant scientific and technological developments taking place during the first half of the nineteenth century and create a unitary vision of science and technology. This book is largely based on Wilson's own writings, and it is the first book-length study of him published in the last 160 years.

This book is essential for researchers and scholars alike interested in Victorian science and technology.

David F. Channell received his Ph.D. in the history of science and technology from Case Western Reserve University. He is on the faculty of the University of Texas at Dallas where he is Professor of History/History of Ideas. He has authored six books on the history of science and technology.

Routledge Studies in the History of Science, Technology and Medicine

For more information about this series, please visit: https://www.routledge.com/Routledge-Studies-in-the-History-of-Science-Technology-and-Medicine/book-series/HISTSCI

George Wilson's Vision of Early Victorian Science and Technology

Unity in Variety

David F. Channell

Routledge
Taylor & Francis Group

NEW YORK AND LONDON

First published 2023
by Routledge
605 Third Avenue, New York, NY 10158

and by Routledge
4 Park Square, Milton Park, Abingdon, Oxon, OX14 4RN

Routledge is an imprint of the Taylor & Francis Group, an informa business

ISBN: 978-1-032-07941-7 (hbk)
ISBN: 978-1-032-07953-0 (pbk)
ISBN: 978-1-003-21221-8 (ebk)

DOI: 10.4324/9781003212218

Typeset in Bembo
by codeMantra

Contents

1   A New Variety

Science and Technology in the First Half of the Nineteenth Century

George Wilson, M.D. (1818–1859), was an important figure in early Victorian science and technology (J. Wilson 1860). He was the first Regius Professor of Technology at the University of Edinburgh (in fact the first anywhere in Great Britain), and he was the founding Director of the Industrial Museum of Scotland (now the National Museum of Scotland). Throughout his career, he lectured and published on a wide range of topics, including the history of science, what we would now call ecology, what we would now call climate change, the prospect of life on other planets, chemistry and poetry, and natural theology. He also argued for the important role of biology in technological development. Wilson's works were very popular. His *Chemistry: An Elementary Text-Book* sold more than 24,000 copies and his popular book *The Five Gateways of Knowledge* sold more than 8,000 copies in two years and was praised by Charles Dickens. Wilson's lectures drew hundreds of students, professionals, artisans, workers, and particularly women (he was an early advocate of educating women in science). He was also somewhat of a poet and published several of his poems. Wilson not only sought to educate people about the significant developments that were taking place in science and technology during the first half of the nineteenth century, but he tried to put these developments into a unitary vision of science and technology by drawing upon religion, transcendental natural history, and Baconian philosophy.

The world that George Wilson was born into was rapidly changing. It was not even yet the Victorian Age, George III was still King. Although Great Britain was in the midst of the Industrial Revolution, there were only a small number of experimental steam-powered locomotives and no railways. People were crossing the Atlantic in sailing ships the same way Columbus had done. The connection between electricity and magnetism had not yet been discovered, so there was no electric telegraph. A large number of the modern chemical elements were still unknown. The planet Neptune had not yet been discovered. Many people believed that the world and all of its creatures was created according to *Genesis* and was only a few thousand years old. The word "scientist" had not yet been coined, and the term technology was almost unknown in Great Britain. University courses in engineering

DOI: 10.4324/9781003212218-1

were almost unknown in British universities. Operations were still being conducted without anesthesia. All of these things would change during the course of Wilson's life.

Chemistry

Some historians have claimed that the period from either 1750 to 1830 or 1800 to 1850 was a "second scientific revolution," or a "multidisciplinary revolution in the 'Baconian sciences'" comprising chemistry, electricity, magnetism, heat, and light (Donovan 1988, 11; Kuhn 1977, chap. 3). For Wilson, one of the most important areas undergoing change was the field of chemistry since he would spend much of his career as a chemist. Many of the changes in chemistry had already taken place before Wilson was born, but he would inherit an entirely new chemistry. The idea of a "chemical revolution" taking place during the late eighteenth and early nineteenth centuries has been much debated by historians of science (Eddy, Mauskopf, and Newman 2014, 1–15; Miller 2004, 34–39; Donovan 1988, 4–12). During the first half of the eighteenth century, chemistry was dominated by a number of theories, including the long-standing Greek idea of the four elements of earth, air, water, and fire. But in addition, there was the theory of essential chemical principles which argued that some material, often water and fire, were the carriers or cause of different chemical qualities (Perrin, 1990, 266; Guerlac 1975, 70; Gillispie 1960, 204). By the beginning of the nineteenth century, chemistry had adopted much of the modern ideas, nomenclature, and techniques that are used today. The reason for these changes, and whether they were actually "revolutionary," is hotly contested among historians. In any case, Wilson was the heir of a "new chemistry" and was often seen as one of the "New Chemists" (Miller 2004, 208). Historians have come to some agreement that the new chemistry was not the result of a crucial discovery of one heroic individual but was the result of a complex web of events that often did not play out in a linear progressive fashion. But any discussion of the "chemical revolution" most often contains some discussion of a series of discoveries and theoretical developments that took place in France and Great Britain in the late eighteenth and early nineteenth centuries.

In 1727, the British chemist and physiologist, Stephen Hales, discovered that "air" rather than always being expansive could sometimes be combined with earth-like substances, especially organic materials and, therefore, become "fixed." In 1755, the Scottish chemist, Joseph Black, discovered that in heating certain substances an air-like material was driven off and labeled it "fixed air," what we call carbon dioxide. Soon after, British chemists began identifying a number of new airs, or gases. In 1765, Henry Cavendish provided a detailed chemical study of "inflammable air" or hydrogen and a short time later Joseph Priestley discovered a number of new gases, including nitric oxide and nitrous oxide, which he called "nitrous airs." Most importantly in 1774, Priestley discovered a new air that he thought was a type of nitrous air but after further experiments in 1775, he realized it was a new type of air (Guerlac 1975, chap. 6). This new air is what we call oxygen, but Priestley interpreted it in terms of the then-accepted theory of phlogiston and called it "dephlogisticated air."

The phlogiston theory arose from the tradition of chemical principles and was used to explain combustion. In the late seventeenth century, the German chemist Johann Becker suggested that all matter was composed of three types of earth-like substances, one of which, *terra pingus*, was the principle of combustion and was lost when a substance was burned. At the beginning of the eighteenth century, another German chemist, Georg Stahl, renamed Becker's *terra pingus* phlogiston from the Greek word for flame. He also expanded the role of phlogiston to the process of calcination, what we call oxidation or rusting, which he saw as a type of combustion. As such chemists began to believe that when a substance burned or underwent calcination, phlogiston was released into the air, but the air could absorb only so much phlogiston so the processes would stop when the air became saturated. This was useful for explaining why a material loses weight after combustion but the fact that metals actually gained weight after calcination became a problem, leading some to argue that phlogiston was lighter than air or actually had negative weight.

The solution of the problem of phlogiston came from Antoine-Laurent Lavoisier, a French chemist who was involved with a number of prerevolutionary government projects such as the study of gunpowder and the purity of Paris's water supplies (Guerlac 1975). His original scientific interest was in geology, but by 1772 he became interested in the chemical composition of the air. As we have seen, British chemists were rapidly identifying a number of new gases or "airs," which led to questions concerning the role of phlogiston in these new gases. Influenced by Hales discovery that air could be "fixed," Lavoisier began experimenting heating metals, expecting that air might be released, but found instead that the metals gained weight as they were calcinated. He came to suspect that the absorption of air was part of all types of calcination and predicted that if true this could bring about a "revolution" in chemistry (Guerlac 1975, chap. 5). At the time he was not sure if the air involved was simply atmospheric air, some part of atmospheric air, or Black's fixed air. By 1775, Priestley had published his discovery of dephlogisticated air that Lavoisier called the purest part of atmospheric air since flames burned brighter in it and it seemed to play an important role in respiration, and Lavoisier came to believe that it was this newly discovered pure air that was playing a role in combustion and calcination (Guerlac 1975, chap. 6).

Lavoisier next turned to a study of acids (Guerlac 1975, chap. 7). A number of chemists believed that all acids share some common principle, but Lavoisier became more interested in finding the actual chemical substance that gave rise to all acids. During his experiments on combustion, he discovered that burning phosphorous and sulfur produced acids that weighed more than the original materials. In 1779 Lavoisier argued that since "eminently respirable air" seemed to be present in most acids that it was the source of acids, and renamed it "oxygen," from the Greek for "begetter of acids" (Guerlac 1975, 93). As it turned out, Lavoisier was wrong about all acids containing oxygen, but the name became universally accepted.

Another result of the new chemistry was a new understanding of the elements. Since the time of the Greeks, water was thought of as one of the fundamental four elements, that is, not decomposable. As we will see in more detail when we discuss George Wilson's biography of Henry Cavendish, during the summer of 1781 Cavendish heard from Priestley that exploding a combination of hydrogen and oxygen left some liquid residue, but Priestley did not follow up on the issue. Cavendish did a series of extensive experiments that gave evidence that hydrogen and water could combine together and form water. The fact that he did not make his results public until 1784 and the fact that he still held to the theory of phlogiston led to the "water controversy" in which claims were made that Henry Cavendish, James Watt, and Lavoisier were each the "discoverers" that water was not an element but a compound of hydrogen and oxygen (Miller 2004). Lavoisier heard of Cavendish's experiments in 1783 before they were made public, and he redid many of the experiments and came to the conclusion that water was not a simple substance but a compound of two gases. While Lavoisier did not discover the fact that water was a compound, he did provide a clear explanation for the chemical composition of water without resorting to the idea of phlogiston (Guerlac 1975, chap. 8). It is sometimes said that British chemists made the discoveries that were then explained by the French chemists.

The fact that a number of Lavoisier's ideas on combustion, acids, and water centered on the role of oxygen led him to seek a reformulation of chemistry (Gillispie 1960, 230). One of his first steps was in 1785 to abandon the concept of phlogiston which he came to see as a tautology since it argued that things burned because they had the principle of burning. Rather than being the source of Lavoisier's "revolution," this was more the result of it. His next step in developing a new chemistry was to reformulate the nomenclature and, thus, the framework of chemistry (Perrin 1990, 273–275; Guerlac 1975, chap. 11; Gillispie 1960, 233–235). Working with Guyton de Morveau, who had been trained as a lawyer, Lavoisier published *Méthode de nomenclature chimique* (*Method of Chemical Nomenclature*) in 1787 bringing about a new order to chemistry. A series of tables were created which listed the simple substances (although later some would be found to be decomposable), and compound substances. In some cases, old names were retained and in other cases newer names were substituted such as oxygen, nitrogen, and hydrogen. Compound substances were labeled with a name indicating what they had in common and then a binomial name indicating something similar to a genus and species (e.g., sulfuric acid, or lead oxide). The nomenclature was not simply a new set of names but a new way to think about chemistry and chemical compounds. This new chemistry was furthered by the translation of the *Nomenclature* into a number of languages, including English and German, and by new publications, such as Lavoisier's *Traité élémentaire de chimie* (*Elementary Text-book on Chemistry*) in 1789 and a new journal, *Annales de chemie*, also in 1789. While there were some complaints by British chemists that using the new nomenclature meant accepting a new theory, that was its purpose. It needs to be noted that Lavoisier's new chemistry was in some cases not so new or revolutionary. For example, his theory of acids and oxygen as the acid former would turn out to be wrong. Also, while Lavoisier did eliminate phlogiston, a fundamental part of his new chemistry was the notion of caloric, originally called the matter of fire, that argued that heat, or temperature, was the result of a physical substance called caloric – an idea that could be seen as quite similar to the doctrine that phlogiston was a substance responsible for burning. Finally, a number of chemists, most importantly Joseph Priestley, continued to support the theory of phlogiston.

While Lavoisier's contribution to the new chemistry ended with his execution during the Terror, there were other contributors to the new chemistry that George Wilson would inherit. Two British chemists made contributions to the new chemistry. First, Humphry Davy, who was Professor of Chemistry at the Royal Institution in London in the early 1800s, showed that two substances that Lavoisier labeled acids did not contain any oxygen. First muriatic acid (hydrochloric acid) did not contain oxygen, and oxymuriatic acid was not even an acid but was the element chlorine. Davy also contributed to the new chemistry by using electricity to isolate a number of new elements, including potassium, sodium, calcium, and magnesium, among others (Ihde 1964, 130–131 and 298–301). The other chemist was John Dalton who developed the theory of chemical atomism. Earlier atomic theories had argued that all atoms were the same, but Dalton argued that each chemical element was associated with a unique atom that differed from the other elements by size and weight. Both Dalton and Davy were opposed to the theory of caloric and argued instead that heat was the motion of particles and not a substance. Wilson was clearly influenced by Dalton since he was the subject of one of his biographical sketches and we will discuss Dalton's ideas in more detail in a later chapter. By the time George Wilson was born and especially by the time he began his medical studies at the University of Edinburgh, chemistry had undergone significant changes. Whether these should be labeled a revolution is something under continuing debate, but those changes certainly represented new chemistry and Wilson was considered one of the "New Chemists."

Natural History

Another area of science that was undergoing rapid change both before and after Wilson was born was the broad area of natural history, especially geology, paleontology, and zoology. These areas were particularly important in Wilson's life since many of his close friends were naturalists and the Industrial Museum of Scotland would be closely associated with the Natural History

Museum at the University of Edinburgh. Older histories of geology and pale-ontology usually contrasted the eighteenth and nineteenth centuries in terms of static models versus progressive, developmental, and eventually evolutionary models, but the actually changes turn out to be much more complex. Some scientists, mostly amateurs or those religiously trained, did hold to the Genesis account and followed Bishop James Ussher's idea that the creation as taken place in 4004 BC (Bowler 2009, 4). Given this young age of the earth, it was easy to believe that not much change had taken place since the creation. The geological world had not changed and the species that we see today had not undergone change but were the result of a special and distinction creation and formed a "great chain of being" going from the simplest forms of life up to human beings (Lovejoy 1936). Once geology began to find evidence and argue for a much older age of the earth, it opened up the possibility that change could take place. In fact, if the earth were millions of years old, it was hard to believe that no change had taken place either in the physical earth or in the species that inhabited it and this leads eventually to Charles Darwin's theory of evolution.

Much of the recent works in the history of geology, zoology, and paleon-tology have shown that the story is complex (Sloan 1990, 295–313; Laudan 1990 314–325). Copernicus's new heliocentric astronomy that resulted in a breakdown of the distinction between the earth and the planets led scientist to begin to speculate how these bodies were formed (Bowler 2009, chap. 2). During the seventeenth and eighteenth centuries, a number of scientists, including the William Whiston in England, Georges-Louis Leclerc, Comte de Buffon and Pierre-Simon Laplace in France, and Immanuel Kant in Germany, speculated that the earth and the other planets formed from a disk of dust that condensed under the force of gravity into the various planets in what became known as the nebular hypothesis. By the middle of the eighteenth century, scientists were beginning to speculate on details of how the earth came from condensed dust into its present form. In 1778, Buffon argued that the earth emerged during six epochs, corresponding to the six days of creation but immensely longer in duration. The earth began as molten, grad-ually cooled, forming a crust, then water vapor condensed forming a great ocean which then retreated exposing land which allowed for the emergence of tropical creatures, and with further cooling the emergence of humans. New empirical studies of the earth, encouraged by the Baconian ideal, and canal building led to the observation that the earth did not have a uniform structure but exhibited layers of different types of earth or strata.

Buffon's theory of the formation of the earth provided a starting point for a debate between geologists that would continue into the middle of the nineteenth century (Bowler 2009, 57–59). Most directly, Buffon's idea of a retreating ocean would become reflected in what became known as the "Neptunist" school of geology, while his idea of heat and a cooling earth would be reflected in what became known as the "Plutonist" or "Vulcanist" school of geology (Laudan 1990, 316–321). During the late eighteenth and early nineteenth centuries, Neptunist theories tended to dominate geology. One of the leading proponents of Neptunist theories was Abraham Gottlob Werner, a professor at the School of Mines in Freiburg, Germany (Jenkins, 2016, 527–557; Bowler 2009, 40–44). Given the spherical shape of the earth, Werner and his followers assumed that at some point the earth had been fluid and postulated a vast ocean covered the entire earth. From this ocean, crystalline rocks, such as granite, were deposited on the ocean floor. As the ocean retreated (they did not give a clear explanation why), it exposed dry land and while rocks such as limestone continued to be chemically precipitated out into the ocean, on land erosion resulting from wind and rain would cause other materials to flow back into the ocean and create new strata of rocks, such as sandstone (Jenkins 2016, 533). This theory allowed Werner to explain a classification system of minerals that he had developed and by postulating that the original ocean floor was uneven and in some locations the ocean levels could rise again, he was able to create an explanatory model that became useful for understanding much of the empirical evidence geologists were discovering such as why some of the oldest rocks, like granite, were found high in the mountains, since they were the first exposed rocks, and younger alluvial rocks, like sandstone formed by erosion, were found in the valleys.

Near the same time that Werner was developing his Neptunist theory, other geologists were putting forward Vulcanist theories that argued that heat was the driving force behind geology (Bowler 2009, 59–620). Some early Vulcanist geologist argued that volcanic activity was behind the system of strata by creating layers of ash. Others argued that the strata were formed under water, but it was the action of volcanoes and earthquakes caused by heat that elevated those rocks to form dry land. The leading Vulcanist, although sometimes labeled a Plutonist, was the Scottish geologist James Hutton. Like the Neptunists, he believed that erosion caused sediments to flow into the oceans causing strata, but he believed that those strata were raised above on ocean by the expansive power of heat in the central core of the earth. These newly exposed strata would again be subject to erosion and wash back into the oceans causing a cyclical or steady-state process unlike the more progressive or directional process of the Neptunists. Also, rather than rocks precipitating out of water, he argued that they were formed by heat and demonstrated that granite was an igneous rock not a sedimentary rock. Hutton's theory was based on his belief in the "uniformity of nature," and on the empirical belief that science should be based on what is observable at the present time (Bowler 2009, 62). As such his theory was similar to Charles Lyell's later theory that came under the term "uniformitarianism."

The debates concerning the origins of the earth became connected with debates over the origin of life on the earth. As noted, there had been a long belief in a great chain of being in which God created all of the species and arranged them in a hierarchical system from the simplest up through humans (Bowler 2009, 62–66). This led to the preformationist theory that all life arose from a series of preexisting "germs," and that each species was fixed.

By the eighteenth century, a more complex version of the origins of life was beginning to emerge (Bowler 2009, 66–81). Buffon argued that a species was a group that shared some "internal mold," but did not clearly explain the nature of that mold. He modified this idea to argue that similar species might share a common ancestor but as members of that original species migrated to different climates and environments they might "degenerate" and diverge from their original family form. He still believed that these families were fixed but could diverge into different "varieties."

In France, a number of scientists who supported materialist beliefs began to provide an alternative to the idea of the fixity of species (Bowler 2009, 81–95). Rather than a "germ" theory of life, they held that life could begin through spontaneous generation from purely material conditions. A leading figure in this movement was Jean-Baptiste Lamarck. Although he first believed that life could only arise from some outward "force" of nature, he came to accept the idea of spontaneous generation. Since he thought only the simplest forms of life could arise through spontaneous generation, he postulated that all higher forms of life developed from these simple forms through the action some type of nervous fluid, similar to electricity, carving out more complex pathways leading to more complex organisms. These pathways were not random but predetermined in terms of some internal plan so the development of the hierarchy of life followed a design or pattern. Unlike Buffon, and later Darwin, Lamarck did not see organisms developing from a common ancestor, rather each type of organism arose from a separate act of spontaneous generation which meant that the simplest organisms are the youngest, while the most complex organisms were the oldest since it would take time for that complexity to emerge. Given that French geologists were arguing that the physical earth was undergoing change, and since Lamarck was firmly against the possibility of extinction, he needed to account for how organisms could adapt to such conditions. This led to his idea of the inheritance of acquired characteristics, which was a minor part of his overall theory but would become the most famous idea for future scientists since it provided an alternative to Darwin's theory of natural selection. According to this theory, the use or disuse of some particular organ as a result of changing environmental conditions would carve new pathways for the nervous fluid, and those new pathways would be inherited by the next generation.

Lamarck's theory of transmutation or transformism was challenged in post-revolutionary France by Georges Cuvier who was the Director of the *Museum d'histoire naturelle* (Bowler 2009, 108–115). Through his studies of comparative anatomy, he concluded that the complex internal structure of most organisms would make changes very difficult. Focusing on internal structure rather than outward appearance, Cuvier was able to develop a new classification system based on function which broke down the linear system used by Lamarck. Cuvier's work in comparative anatomy led him to begin a detailed study of fossils. Although he rejected a theory of transformation of organisms, his work in paleontology would end up providing evidence in support of transformism as well as a new theory of geology. As early as the seventeenth century scientists began to recognize that fossils, rather than being simple natural rock formations, were the remnants of living organisms whose existence they often attributed to the flood of Noah (Bowler 2009, 35–38). The fact that many of the fossils that Cuvier studied did not correspond to any currently living organism led him to the radical conclusion that groups of animals had gone extinct throughout the past. Based on Werner's Neptunist theory that the earth was composed of a series of different types of rock formations, Cuvier began to associate different types of fossils with different strata and was able to place them in a geological sequence (Bowler, 2009 111–115). While Cuvier, like Werner, saw water as the primary force shaping the earth, unlike Werner he thought that the geological changes were brought about by sudden catastrophic floods and developed a theory that became labeled catastrophism, but he did not associate any of the catastrophes with the biblical flood.

Older histories of geology and evolution tended to see Darwin's theory of evolution as the culmination of nineteenth century natural history and, therefore, downplayed the role of Neptunism and Lamarckianism in early Victorian science, but recent historians have provided a more complex view. First, although Darwin took Charles Lyell's uniformitarian work *Principles of Geology* on his voyage on the *H.M.S. Beagle*, he had also studied with Adam Sedgwick, a leading catastrophist, while at Cambridge. His own grandfather Erasmus Darwin's work *Zoonomia* had a number of elements in common with those later put forward by Lamarck and near the end of his life Darwin seems to have become open to certain neo-Lamarckian ideas (Bowler 2009, 236). Some recent studies have shown that both Neptunism and Lamarckianism continued to be actively debated in Edinburgh, where Darwin attended medical school, into the 1850s (Jenkins 2016; Secord 1991). Bill Jenkins and James Secord make an argument that Robert Jameson, Professor of Natural History at the University of Edinburgh and a student of Werner, managed to combine Neptunist geology with a progressive or transformist view of life by drawing on Lamarckian ideas. This contradicted the older histories that saw transformist theories leading to evolution and Neptunism as an out-of-date theory (Jenkins 2016, 529–530). While Lamarck argued that the progressive development of life was primarily the result of some innate drive toward complexity and only secondarily the result of the environment, Jameson argued that the transformations of life were driven by climate change that was the result of the cooling of the earth from its original molten state which he also associated with the Neptunist idea of the retreat of the oceans. Jenkins quotes Jameson as saying: "As the water diminished, it appears to have become gradually more fitted for the support of animals and vegetables, as we find them increasing in number, variety and perfection" (Jenkins 2016, 535). James Secord finds additional evidence in an article entitled "Observations on the Nature and Importance of Geology" which was published in 1826 in the *Edinburgh New Philosophical Journal* which was edited by Jameson and became a venue for discussions of Wernerian geology (Secord 1991). Secord makes a convincing argument that this anonymous article was in fact written by Jameson and says: "For the author of the 'Observations,' this progression [from sea creatures to land animals to humans] of life is best explained through transmutation. Lamarck's theory is the logical consequence of Werner's" (Secord 1991, 9). In agreement, Jenkins says: "Directional change in the surface of the globe, of the kind that is integral to the Wernerian model of the history of the earth, is therefore put to the center of this [the 'Observations'] theory of transmutation" (Jenkins 2016, 539). This evidence seems to indicate that both Neptunism and Lamarkianism were alive and well, at least in Scotland, well into the nineteenth century. At the same time in England, catastrophism, as put forward by William Buckland and Adam Sedgwick, was becoming the dominating geological theory (Bowler, 2009 115–120). Part of this may have had to do with the influence of William Paley's argument from design and the popularity of natural theology in England. As we will see in a later chapter, natural theology was seen as protecting scientists from charges of atheism and radical thought. As such, Cuvier's catastrophism, with its molten earth and subsequent floods and earthquakes, seemed to fit more easily with a biblical account of the earth's creation.

As we will see in the next chapter, Edinburgh also became a center for what would be labeled philosophical or transcendental naturalism through the teachings of Robert Knox and Robert Jameson (Rehbock 1983, chaps. 2 and 3). Transcendentalism had its roots in the philosophy of Immanuel Kant who argued that the human mind played an active role, through what he called categories, in organizing sense experience and, therefore, transcended the world of physical appearance (Rehbock 1983, 16–17). Kant's transcendental idealism influenced a number of other German philosophers and scientists, such as Friedrich Wilhelm Joseph von Schelling and Johann Gottfried Herder, among others, who developed a theory known as *Naturphilosophie* that became connected with Romanticism. Kant's focus on the role of the human mind in understanding the world led the *Naturphilosophen* to view "the universe as a manifestation of the divine Mind and encouraged the search for an underlying unity in nature" (Bowler 2009, 121). A leading example of this approach can be found in the work of Johann Wolfgang von Goethe. Influenced by both Kant and Plato, he sought to find a basic unity in the plant world and thought of "various parts of the flowering plant as successive modifications or metamorphoses of the basic form, an idealized primal leaf. Beyond that, he regarded all flowering plants as variations on an ideal plant archetype, or *Urpflanze*" (Rehbock 1983, 19).

One important element of this idealistic or transcendental naturalism was the idea that embryology could provide insights into the development of life on the earth (Bowler 2009, 121–124). The idea that living things followed some "unity of plan" led a number of naturalists, including C.F. Kielmeyer, F. Meckel, Lorenz Oken, and Louis Agassiz, to put forward a law of parallelism which argued that the development of a human embryo paralleled the development of other forms of life on the planet (ontogeny recapitulates phylogeny). That is, the human embryo went through a stage similar to a fish, then a reptile and then a mammal (some people thought they saw evidence of gills in early human embryos). The supporters of this theory did not believe that one species was transmuted into another species, but that all of the species develop in a linear progression with some species ending their development before others (e.g., mammals would pass through a fish and a reptile stage, while reptiles would pass only through a fish stage). While not evolutionary in the Darwinian sense, such a theory was developmental or progressive. The law of parallelism was challenged by Karl Ernst von Baer who argued that at early stages of development, embryos of different species were difficult to distinguish from one another and they do not pass through other stages in a linear way, rather they radiate from a common point of origin by becoming more specialized and more complex. While von Baer's theory was nonprogressive, it was still based on the idea of some archetype or unity of plan and his idea of branching development provided an idea that would become useful in Darwinian evolution (Bowler 2009, 124). An idea of branching development similar to von Baer's would become an important element of the transcendental naturalists at Edinburgh led by Robert Knox (Bowler 2009, 124–126; Rehbock, 1983, chap. 3).

While the majority of theories of natural history in the first half of the nineteenth century, even the idealists or transcendental theories, held that the earth and the living things inhabiting the earth had undergone some type of change over time, Charles Lyell resurrected a more steady-state, or at least cyclical, theory of natural history similar to one put forward earlier in the century by James Hutton (Bowler 2009, 129–133). This is somewhat ironic since Lyell, although never accepting Darwin's evolution, would have one of the biggest impacts on the development of Darwin's theory. As noted earlier, Darwin took a copy of Lyell's *Principles of Geology* with him on his voyage of the *Beagle*. As a Unitarian, Lyell became opposed to catastrophism, especially any identification of the flood of Noah as one of the catastrophic episodes that shaped the earth. Instead, Lyell based his geology on a methodology labeled "actualism," which argued that explanations of geology should be based on events and forces that are observable taking place during the present time thus eliminating the use of miraculous events like worldwide floods, enormous earthquakes, or super-volcanoes. Lyell based this methodology on the belief that nature acted in a uniform way, and it, thus, came to be known as "uniformitarianism," but he was somewhat vague on what aspect of nature was uniform – was it the laws of nature, the kind of causes operating on the earth, or the intensity of those causes (Laudan 1990, 321). Most geologists accepted that the laws governing nature had not changed but given the evidence that the earth had undergone significant cooling, it was harder to accept that fact that the kind of causes and their intensity had not changed. Even Lyell had to accept that in the case of the origin of species, the kind of causes might have changed since he believed that the species were divinely created.

A consequence of Lyell's idea of uniformitarianism was that the time frame of geology had to be dramatically expanded. If mountains and valleys were the result of the types of earthquakes and erosion that we observe on the earth today, those forces would have to have been acting over a great period of time. This steady-state model of the slow rise of mountains and their subsequent slow eroding away did not mean that the earth remained unchanged throughout history but that the changes taking place were slow gradual changes, an idea that Darwin found to be central to his theory of evolution. Another important element of Lyell's geology that would be important to Darwin was that slow geological changes could affect the environment which, in turn, could affect life (Rehbock 1983, 153–156). As a result, the earth would not necessarily be unchanging. For example, at some periods of time or in some locations, the environmental conditions might favor reptiles over mammals and at other periods and places it would favor mammals over reptiles, but the development of life would not be progressive (Bowler 2009, 133–134). While mammals dominate in the present age, it could be possible for an age of dinosaurs to return in the future and that higher levels of mammals might have existed at all times in the past, but that their fossils simply have not yet been discovered. This argument for the incompleteness of the fossil record would become used by Darwin to support his theory. The changes in natural history taking place in the first half of the nineteenth century would culminate with the unifying theory of Darwinian evolution, but Darwin's work was not published until 1859, two days after Wilson died, and did not have a direct influence on him, but it will be discussed in the final chapter.

Natural Philosophy

During the first half of the nineteenth century, natural philosophy underwent significant changes that culminated in the development of the laws of thermodynamics, at the center of which was the new unifying concept of energy (Smith 1998; Smith and Wise 1989; Harman 1982; Cardwell 1971; Kuhn 1959). Closely associated with the work on the correlation of forces that would lead to the formulation of the conservation of energy was the development of the concept of electromagnetic fields that would culminate in James Clerk Maxwell's set of equations governing electricity and magnetism (Harman 1982, chap. 4; Doran 1975; Williams 1971; Hesse 1965; Gillispie 1960). Natural philosophy in the nineteenth century was still dominated by the theories of Isaac Newton, but several issues had been left unresolved. The biggest issue was the nature of gravitational force and its relationship to matter. In the *Principia*, Newton had famously said that he did not "feign" a hypothesis concerning the nature of gravity and action at a distance. Later in the *Opticks*, he suggested that gravitational action at a distance might be explained by hypothesizing that the universe was filled with a fine subtle fluid, labeled the aether, that had a higher density inside of matter than in space (Channell 1991, 23–26). But this only pushed the question to a lower level since his aether was composed of fine particles that filled the universe by being self-repelling. So, in order to explain the attractive action at a distance of gravity, one had to assume the action of repulsive action at a distance of the particles of aether. During the eighteenth century, a number of scientists extended Newton's idea of an ether and tried to explain attractive and repulsive forces as being associated with some subtle, imponderable fluids, such as an electric fluid that would explain the attractive and repulsive properties of electricity or a caloric fluid that would explain the expansive nature of heat, although the exact nature of these fluids, except their weightlessness, was never addressed (Schofield 1970). A problem with assuming that the attractive and repulsive nature of force was simply the result of the actions of some material substances was that it opened to door to materialism and atheism since there would no longer be any need for a non-material spirit guiding the universe. This was a particularly popular interpretation in France.

During the second half of the eighteenth century, the theories of the Croatian natural philosopher Roger Joseph Boscovich provided the possibility of a new interpretation of the relationship between force and matter that would become increasingly influential during the nineteenth century, especially in England and Scotland (Olson 1975, 98–106; Williams 1971, 73–80; Scholfield 1970, 236–241). In his *Theoria Philosophiae Naturalis* (1758, 1763), Boscovich put forward a non-materialistic theory of matter. Rather than employing physical atoms, for him matter was composed of a series of dimensionless points that served as centers of attractive and repulsive forces. Near the points, the forces were highly repulsive giving the impression of hard-billiard ball like atoms, but that was simply the effect of the existence of highly repulsive forces. Further away from the point centers, the repulsive forces became attractive forces that followed Newton's law of gravitation. Boscovich saw this model as explaining a problem associated the atomic collisions (Olson 1975, 99–100; Williams 1971, 73–74). One could not explain atomic collisions as if they were like billiard balls since the elasticity of billiard balls is the result of them being composed of small atoms that allow the billiard balls to deform, but true material atoms cannot have internal parts and, therefore, at impact the atoms would have to have their original velocity and their rebound velocity. Having two velocities was a logical contradiction. His radical solution was to eliminate matter and simply rely on continuous space-filling forces. The fact that Boscovich's forces went from attractive to repulsive in a continuous way allowed for colliding "atoms" to experience a continuous increase in repulsion, and the problem of two velocities was eliminated. It should be noted that beyond the highly repulsive center and the longer distance attraction of gravity, the forces could oscillate back and forth from attractive to repulsive and since ordinary matter is actually composed of a complex set of these point force centers, there can be very complex patterns of interaction with other ordinary matter. Also, since everything is fundamentally simply a set of forces, those patterns of forces can transform from one type of phenomenon to another type.

By the beginning of the nineteenth century, a number of natural philosophers, especially in Germany and Great Britain, were coming to reject the materialism of the French and coming to accept non-materialistic theories. In Scotland, Dugald Stewart became a strong supporter of Boscovich's idea, seeing them as supporting many of the ideas of Common Sense philosophy (Olson 1975, 98–106). In Germany, Immanuel Kant was developing ideas that were similar to Boscovich's (Williams 1971 60–73). Kant in his *Critique of Pure Reason* tried to address the same problems raised by Hume's skeptical philosophy that had troubled the Common Sense philosophers. Like the Common Sense philosophers, Kant argued that the basic source of our knowledge came from our senses and also like the Common Sense philosophers, that information was not simply recorded in a passive way in the mind but rather shaped by structures (intuitions and categories) in the mind. Our senses cannot provide knowledge of the world as a thing-in-itself (*Ding an sich*) but can only provide what comes through our senses, that is, our knowledge of matter as impenetrable (repulsion) or being attractive (gravity). This led Kant to put forward in his *Metaphysical Foundations of Natural Science* (1786) a natural philosophy that instead of depending upon matter was based on attractive and repulsive forces.

Kant's ideas were pushed even further by a group of German philosophers, many of whom were linked to the Romantic movement, and who put forward a philosophy labeled *Naturphilosophie* (Rehbock 1983, 16–21). A leading figure in this movement was Friedrich Wilhelm Joseph von Schelling. His philosophy was based on a number of principles, including the idea that the universe was a reflection of the mind of God and that humans through their minds could discover the transcendental reality of the universe. This meant that all of the natural world was alive, functioning like a purposeful organism, in which non-material or spiritual causes were just as important as "material" causes. In fact, following Kant, the *Naturphilosophen* saw the world as only composed of non-material polar forces (attractive and repulsive) which through continual tensions and conflicts gave rise to higher level and more complex forces. It then became of role of natural philosophy to discover the underlying unity of nature, and this led a number of early nineteenth-century natural philosophers to focus on the connections and correlations between what appeared on the surface, to be apparently different types of phenomena. This, in turn, would lead to the discovery that the concept of energy, that came to be used instead of force, was something that was conserved in all scientific process and came to serve as a new "reform programme," for nineteenth-century natural philosophy (Smith 1998, 7).

Even before the use of the term energy, the idea that a unity and connection existed between natural phenomena was not just the focus of attention of natural philosophers but was being brought to the attention of the general public. In 1834, Mary Somerville, a Scottish scientist who began studying mathematics but extended her interests into astronomy, chemistry, geography, electricity, and magnetism, wrote a book entitled *On the Connexion of the Physical Sciences* (Somerville 1849). The book was widely popular, and it was the largest selling scientific book until Darwin published *On the Origin of Species* in 1859. Somerville gives descriptions of the connections between a wide variety of other phenomena that were currently being discovered by researchers. A recent reviewer claims that her most original work involves Michael Faraday's work and that these sections "clearly predict the connection between all electromagnetic phenomena established in four equations a generation later by physicist James Clerk Maxwell" (Holmes 2014, 432). Along with new work on electromagnetism, Somerville book reports and explains new work being done on the connection between heat and light, the connection between waves in water and air, the connection between infrared and ultraviolet light, the effect of light on chemicals in photographic plates, and the global connections of scientific research programs involving exploring the arctic, the atmosphere in balloons, and Humboldt's explorations in South America (Holmes 2014, 432–433).

As Crosbie Smith has argued, the conservation of energy was not something that was "discovered" like a new planet, but it was an idea that was constructed over a period of time through the work of a number of researchers in several locations like those reported by Somerville (Smith 1998, 12). In an often-cited article, Thomas Kuhn outlined the wide-ranging early nineteenth-century research that eventually came together in the formulation of the idea of the conservation of energy (Kuhn 1959). Rather than thinking in terms of simultaneous discovery, Yehuda Elkana argues that different researchers in different locations were working on different sets of problems whose solutions turned out to be related to one another (Smith 1998, 11). In any case two areas of research seemed to play an important role in leading scientists to think that many different types of natural phenomena could be converted into one another and that a way to explain that interconvertibility was through the idea of the conservation of energy.

One of the most important discoveries of the interrelationship between two apparently different phenomena was the discovery of electromagnetism (the fact that a moving electric current could create a magnetic effect) by the Danish physicist Hans Christian Oersted in the winter of 1819–1820, and the discovery of electromagnetic induction (the fact that the motion between a wire and a magnet can induce an electric current into the wire) by the English physicist Michael Faraday in 1831 (Williams 1971, 137–144 and 169–183). As we will see later in this chapter, not only were these fundamental scientific discoveries but they would also lead to new systems of communication and power. Electric and magnetic phenomena had been observed for centuries and while they exhibited some tantalizingly similar properties (such as like magnetic poles repelling one another and unlike poles attracting one another, while like electric charges repelled one another and unlike charges attracted one another), they seemed to be two different phenomena. But, as L. Pearce Williams has famously argued both Oersted and Faraday, influenced by the works of Boscovich, Kant, and the *Naturphilosophen*, concluded that if all phenomena could be reduced to attractive and repulsive forces, then there might be connection between electricity and magnetism. In fact, Oersted believed that there was a common force that created the phenomena of heat, light electricity, and magnetism and these could be converted into one another (Williams 1971, 138). Oersted famously "accidentally" discovered electromagnetism when during a lecture at the University of Copenhagen he noticed that a compass needle was deflected when he connected a wire to a battery. Of course, this was not accidental since he had been searching for a relationship between electricity and magnetism since at least 1806. Rather, his discovery is a perfect example of Louis Pasteur's statement that "Chance favors the prepared mind." On the other hand, Faraday seemed to have worked through a systematic series of theories, driven by the ideas of Boscovich, that resulted in his discovery that if you moved a magnet in and out of a loop of wire, an electromotive force was induced into the wire. Faraday explained electromagnetic induction by using the idea of lines of force that could spread through space; an idea that was influenced by his reading of Boscovich (Williams 1971, chap. 7, esp.281). Later, these lines of force would serve as his basis for the concept of electromagnetic fields that would be described later by Maxwell's equations (Williams 1971, 506–513).

Another leading set of developments that led to the formulation of the conservation of energy occurred in the area of the study of heat. In fact, the first law of thermodynamics is the conservation of energy (Channell 2019, 100–108; Cardwell 1971). During the eighteenth century, it had been widely accepted that heat was an actual fluid-like material (Gillispie 1960, 362–370). Ever since the Greeks, fire had been thought to be one of the four basic elements and by the eighteenth century heat was thought to be one of the imponderable fluids, called caloric, which produced the sensible effect of heat. Although caloric did not have any weight, as a fluid it was thought to have extension, to act repulsively and to be conserved. Although the steam engine had been invented in 1712, no one exactly understood the scientific principles behind it. Since condensing steam was used to create a vacuum which allowed the weight of the atmosphere to move the piston, it was widely thought to be a pressure engine. By the early nineteenth century, the Industrial Revolution was in full force, and steam engines were being widely used in manufacturing and transportation. This led to questions as to whether existing designs were the most efficient or if engines could be made more efficient. It was then becoming realized that the power of the steam engine was derived from heat, and this led to a new interest in the nature of heat. A new understanding of the role of heat in the steam engine emerged in 1824 with the publication of Said Carnot's *Réflexions sur la puissance mortice du feu* (*Reflections on the Motive Power of Heat*) (Channell 2019, 98–100). Still using the old caloric theory of heat, Carnot was able to argue that a perfect heat engine would be one in which heat fell from a high temperature to a low temperature in order to do work, and in a perfect engine there would be no useless flow of heat, rather heat would be used to cause some substance to expand (Cardwell 1971, 193). While Carnot's work led to a new understanding of heat engines, it still relied on the idea of the conservation of heat, and the development of a science of thermodynamics would have to take into account that heat was not simply flowing from a hot body to a colder body but in the process heat was being transformed into work.

Several people played a role in the discovery that heat and work were interconvertible (Channell 2019, 100–104). In 1798, the American ex-patriot Count Rumford (Benjamin Thompson) undermined the caloric theory of heat by showing that in boring out a cannon, an apparently inexhaustible amount of heat could be generated through friction and that no physical change had taken place in the material of the cannon. But it was not until the 1840s that the German physician Julius Robert Mayer, and the British scientist, James Prescott Joule, would establish a mechanical theory of heat which would lead to the law of the conservation of energy. As a ship's surgeon in the East Indies, Mayer observed that venous blood in the tropics was much brighter red than in cooler climates. He attributed this to the fact that the tropical climate required less oxidation to keep a body warm. The additional observation that physical activity could raise the internal heat in an animal led him to argue that there must be some connection between heat and work. Influenced by *Naturphilosophie's* idea that *Kraft* (force) was not a property of matter but an independent phenomenon, and was indestructible, Mayer argued that heat and work were interconvertible. This has led some to claim that Mayer had formulated the idea of the conservation of energy, but others argue that his concept of *Kraft* needs to be distinguished from what would become the concept of energy (Smith 1998, 74). Also, Mayer argued that heat was a particular type of force rather than a type of motion and, therefore, did not put forward a dynamical or kinetic theory of heat. On the other hand, he did conduct some experiments measuring the relationship of a column of mercury and the heat generated during the compression of a gas and concluded one kilocalorie of heat could be generated by 1 kg falling 365 m (Cardwell 1971, 230).

While some credit Mayer for developing the idea of a mechanical equivalent of heat, many credit Joule for providing the scientific evidence for it (Channell 2019, 102–104). Joule was initially interested in studying electric motors as a potential new source of power but concluded that the chemical consumption of zinc in electric batteries was much more expensive than the consumption of coal in a steam engine. But his studies of electric motors led him to a study of the relationship between heat and work. While it was clear that the heat in an electric circuit powered by a battery arose from chemical action, he also discovered that a magneto (a device that generated electricity by rotating a coil in a magnetic field) also led to heat in a circuit. This led him to see that heat could be generated by mechanical activity and was not a substance like caloric. Joule expanded his experiments beyond electric motors and showed that mechanical action could be converted into heat by measuring the amount of heat generated in a cylinder of water by a set of paddle wheels attached to a falling weight. This led him to argue that there was a fixed mechanical equivalent of heat and that the interconvertibility of heat and work could only take place if heat were some type of motion (Smith 1998, 72; Cardwell 1995, 311).

By the late 1840s, the interconvertibility of heat and work was becoming generalized into the law of the conservation of energy which would become a unifying scientific principle (Channell 2019, 103–107). In a 1847 paper entitled "On the Constancy of Force" (*Über die Erhaltung der Kräft*), Hermann von Helmholtz, one of Germany's leading scientists, provided a theoretical and mathematical proof of what would become the conservation of energy (the term energy was not yet in wide use so he used the term force). Helmholtz's early studies in physiology convinced him that animal heat could be explained by the oxidation of organic substances in the lungs and muscles, and this, in turn, was based on the "constancy of force," by which he meant forces were indestructible but could be transformed into one another. Helmholtz further believed that this constancy of force did not simply govern living things but was a general principle that governed all of nature. To prove this, he conducted a mathematical investigation in which he was able to show that in any system of material objects governed by Newtonian attractive and repulsive forces whose intensity depended on the distance between them, the change in what would come to be called kinetic energy would be equal to the change in what would come to be called potential energy and, therefore, the total energy would be conserved and this principle was applicable to all of nature (Channell 2019, 104).

What Crosbie Smith has called the "science of energy" was formulated during the early 1850s primarily by the Scottish scientists and engineers W.J.M. Rankine, William Thomson (later Lord Kelvin), James Clerk Maxwell, and P.G. Tait, and the German scientist and engineer Rudolf Clausius (Smith 1998, chap. 1). Their major problem was to reconcile the work of Sadi Carnot, whose theory rested on a caloric theory of heat, with the discoveries of J.P. Joule, whose work rested on a dynamical theory of heat. The result of their work was to realize that heat was not always conserved, but in the process of heat being transformed into work some of that heat was dissipated and unable to do useful work but it had not been destroyed. This research led to two universal laws of thermodynamics using two new terms – energy and entropy (a measure of the heat lost through dissipation). The first law stated that the total "energy of the universe is a constant," and the second law stated that "the entropy of the universe tends to a maximum" (Smith 1998, 168). As a result, the science of energy provided a new unitary framework for understanding the natural world.

Technology

At the same time, that natural philosophy was undergoing significant transformations, during the late eighteenth and early nineteenth centuries, large-scale changes were taking place in technology as a result of the Industrial Revolution (Brose 1998; Cardwell 1995; Deane 1969; Landes 1969). Historians do not all agree on the dates of the Industrial Revolution but there is a consensus around the period 1750–1850, although nothing really important happened on either of those dates. There have also been debates concerning the term "Industrial Revolution," which was popularized by Arnold Toynbee in a set of lectures in 1884 (Toynbee 1887). Can something that took place over a period of one hundred years be considered a revolution? Given that at Toynbee's time little work had been done on the history of technology many of the technological developments appeared to simply burst on the scene, while today we would interpret many of the developments as having roots at least 200 or 300 years earlier. While there certainly were evolutionary aspects of the Industrial Revolution if looked at over one hundred years, there seem to be revolutionary aspects of what was taking place. Before 1750, most structures and machines were made of wood or stone, while after 1850 machines and large structures like bridges and ocean-going ships were made from iron or steel which relied on coal (coke) rather than wood (charcoal) for their production and, therefore, became much cheaper. Before 1750, most manufactured products, especially textiles, were made by craftsmen using simple tools, like spinning wheels and hand-looms, but after 1850 most of the manufactured products were being produced in factories by water- or steam-powered machines. Before 1750, the basic source of power was animal or waterpower, while after 1850 steam power became dominant (and later electric power produced by steam). Before 1750, land transportation was slow, depending on canal boats or horses, but by 1850 goods and people were being transported by railways traveling at the speed of present-day automobiles. Finally, before 1750 most communication depended on physically carrying a message from place to place, which could take days or weeks, but after 1850 the electric telegraph allowed for almost instantaneous communications.

The roots of the Industrial Revolution can be traced to a number of events. One of the significant events was the demographic change that was taking place in Europe, but particularly in Great Britain (Deane 1969, chap. 2). Before 1740, the population growth of Great Britain had been relatively small or even stagnant and had been kept in check by plagues, warfare, and other catastrophic events, but beginning in 1740 those catastrophic events began to decline and the population, while still growing at a relatively small rate, continued to grow and importantly were not reversed by catastrophes but continued to grow faster through the early nineteenth century reaching a rate of almost sixteen percent. There has been a great deal of debate concerning the reasons for this population growth. Some have suggested that a decline in the plague because of a shift in the rat population that carried the fleas that spread the plague may have reduced the death rate, especially the infant mortality rate. Others have suggested that earlier marriages, because of a better economy, led to a rise in the birth rate which could have also been tied to a decline in infant mortality allowing more people to reach child-bearing age. There were also several years of good harvests and products from the new world, such as potatoes, corn and fruits may have provided a wider source of foods. The importance of this population growth is that it provided new local markets for British products, it provided incentives for growth in production, and it provided a source of labor that would be needed for the Industrial Revolution.

The population growth after 1740 could have stifled the development of the Industrial Revolution if it had not been accompanied by what some have called an "agricultural revolution" that allowed the increasing population to be fed (Deane 1969, chap. 3). Beginning in the seventeenth century, the British government passed a number of Enclosure Acts, fencing off the common areas that had been shared by small farmers since medieval times. While very socially disruptive, the Acts lead to what Oliver Goldsmith labeled the "deserted villages," but they also led to larger scale farms and increased the efficiency of agriculture through and economy of scale. By the eighteenth-century inventers, such as Jethro Tull, were creating new agricultural implements, such as the seed drill and the horse hoe, which increased the productivity of agriculture. Other developments, such as Thomas Coke of Norfolk's new system of crop rotation, allowed for much better use of the existing land. New crops, such as the turnip which served as winter fodder, were introduced into Britain. Also, through artificial selection the size and quality of livestock increased. In all of this, there was great support from the British government and the upper classes who were landowners and derived much of their income from agriculture. King George III was known as "Farmer George" and experimented with new farming techniques at Windsor. Aside from allowing a growing population to be fed, the agricultural revolution also led to increases in incomes for a large number of the population, which provided capital needed for industrialization and allowed the public increased purchasing power to buy the new products being turned out by the industrial factories (Deane 1969, 50).

A final trend that helped create the Industrial Revolution was what might be called a "commercial revolution" (Deane 1969, chap. 4). The mercantile system of international trade that arose after Europeans came into contact with America, Africa, and Asia required new organizational structures and new institutions that would be essential in managing an Industrial Revolution. Great Britain, being a relatively small island country with limited valuable natural resources, beyond wool, was particularly dependent on international trade for natural resources and for markets for manufactured products. While something like banking may be found in the ancient Middle East, in ancient Greece and Rome, and in ancient India and China, the more modern concept of banks began to emerge in Italy and later the Dutch Republic during the fifteenth and sixteenth centuries. In England, the Bank of England had been founded in 1694 (Deane 1969, chap. 11). Aside from playing a crucial role in determining the money supply, banks were important in the financing of the mercantile trade by issuing letters of credit, and the British banking system was one of the most advanced in Europe (Landes 1969, 74–75). Mercantile trade was also risky and dangerous and in response, new insurance companies began to arise to distribute the risk. One of the earliest emerged in 1686 from Lloyd's Coffee House near the London dockyards where merchants, shipowners, and sailors came together to exchange information and to insure maritime ventures. This system of international trade was complex and in order to manage it, large joint-stock companies, like the East India Company (founded in 1600) and the Hudson Bay Company (1670), emerged as semigovernmental, semi-military institutions that governed an overseas trade. The managerial skills developed in running these large organizations, such as accounting practices, record keeping, contracts, quality control, and issuing reports would become important in running the new factories that arose during the Industrial Revolution. Finally, the English Patent Law played an important role in the development of the Industrial Revolution (Zukerfeld 2014; Mokyr 2009; MacLeod 1988). While what might be called modern patents were issued as early as 1450 in Venice, most of the early patents were simply a form of royal patronage, but this changed with British Statute of Monopolies passed in 1624 in order to limit the power of monarchy. The resulting new patent system rewarded inventor with a fourteen-year monopoly. Joel Mokyr has argued that without the possibility of becoming wealthy because of a fourteen-year monopoly, there would have been much less inventive activity during the Industrial Revolution (Mokyr 2009, 355).

As a result of the population growth, the agricultural revolution that helped to feed that growth, and the commercial revolution that helped to finance that growth, there was increasing pressure to find new ways and new technologies to make production and manufacturing more efficient. This led to a number of new technological developments that arose in the second half of the eighteenth century that would become the "leading sectors" of the Industrial Revolution. Most of these developments had little dependence on science, as would developments in the nineteenth century, and some of them originated early in eighteenth century and it took them some time before they had widespread impact, providing some evidence that the Industrial Revolution was just as much evolutionary as revolutionary.

One of the most important developments in driving the Industrial Revolution was the change taking place in the iron industry (Deane 1969, chap. 7). While iron had been known and used for tools, weapons, and other technologies since the time of the ancient Greeks, it has always been expensive to produce on a large scale since it required large amounts of charcoal to transform iron ore (iron oxides) into cast iron and charcoal was derived from wood which was often in short supply and needed for things such as shipbuilding. In fact, during the period before the Industrial Revolution, England was becoming deforested and much of the iron making had to be done in Scotland. This added to its cost since the iron ore, which was more plentiful in England, had to be shipped to Scotland, which had more reserves of wood. In 1709, in the small town of Coalbrookdale, near the Welsh border, Abraham Darby perfected a method to produce cast iron using coke (burnt coal). While coke and charcoal were mostly carbon, coal often had other impurities, such as sulfur and potassium which made the cast iron brittle. Darby found a way to eliminate the impurities in coal and produce a useable form of cast iron. Since England had great reserves of coal, this dramatically reduced the price of iron, but it took until the second half of the eighteenth century before Darby's process became widespread and was able to produce cheap cast iron for steam engines and large structures like the iron-bridge which was built by Abraham Darby III and opened in 1781. Without a cheap source of iron, steam engines and railways would not have been economically feasible.

Another new invention that helped spur the Industrial Revolution was the invention of the steam engine (Cardwell 1995, 121–134 and 156–168). Before the Industrial Revolution, most manufacturing in Britain depended on waterpower, or in a few cases wind power, but both these sources of power were restricted to certain geographical locations and were not always reliable, such as during droughts, freezing weather, still winds, or storms. Using the scientific discoveries made during the seventeenth century that the atmosphere had weight and the condensing steam could cause a vacuum, Thomas Newcomen, an English inventor, combined the two ideas and produced what many consider the first workable steam engine in 1712 which was used to pump out tin mines in Cornwall. Like Darby's new method of producing iron, Newcomen's early steam engines had limited impact on manufacturing since they were primarily used as pumping engines. It was not until 1769, when James Watt patented a number of improvements to the Newcomen engine, making it much more efficient, and making it double-acting so it could power machinery, that steam power began to become widespread and serve as a power source for the Industrial Revolution.

One set of new technologies that could be considered truly revolutionary were the new textile machines that were invented in the second half of the eighteenth century (Cardwell 1995, 141–147; Deane 1969, chap. 6). Before the Industrial Revolution, textiles had been produced by hand for hundreds of years. Some fibrous material, such as wool of cotton, was combed in a process called carding so the fibers were all aligned. This was done using something like a simple comb. Then the material was spun into thread using a simple spinning wheel, and finally the threads were woven into cloth using a hand-loom. With the increases in population, this system could not keep up with demand, and the shift from rural areas to cities meant that the people living in cities could no longer produce their own textiles as had been done in rural areas. Beginning in 1769, a series of inventions transformed the textile industry so by 1785, every element of textile production that had been done by hand was now being done by water-powered machines. This allowed Great Britain to become an economic power.

As the Industrial Revolution continued into the nineteenth century, new inventions began to take advantage of the changes that were taking place in science during that period. For example, the chemical revolution led to significant new chemical industries (Channell 2017, 30–40; Beer 1959; Clow and Clow 1952). The German chemist Justus von Liebig stimulated practical applications of chemistry in both Germany and in Great Britain. While making important chemical discoveries, Liebig also played an important role in applying the new organic chemistry to practical areas such as pharmaceuticals and agriculture, especially through his book *Chemistry as Applied to Agriculture*. Through trips to England, he helped to encourage the creation of the Royal College of Chemistry. William Perkins, who was a student at the school, discovered the first aniline dye that helped to begin a large new chemical dye industry throughout Europe, but especially in Germany. Chemical dyes were a great improvement over vegetable dyes since they had the potential of producing a larger number of colors that were more vibrant and more fade-resistant and their production could be standardized. Because of the great commercial success of chemical dyes, chemists began applying the new techniques and knowledge that had come from the chemical revolution to systematically search of new dyes. In the process, they discovered not only new dyes but also a range of new pharmaceuticals. The great success of chemical dyes and pharmaceuticals led many chemical companies to expand their research outside of organic chemistry into the production of alkalis, acids, fertilizers, and explosives.

Two of the most important inventions to come out of Industrial Revolution during the first half of the nineteenth century were the invention of the railway and the telegraph, both of which drew on new scientific knowledge that had been developed during that period. Both inventions would have a profound effect on Victorian life and help to unify the Victorian world. The Newcomen and later Watt steam engines were very large, heavy machines that were not suitable for powering land transportation. They would find a role in powering river boats and ocean-going ships but because of their weight, they were not powerful enough to move vehicles on land and the only way to make them more powerful was to make them bigger since they depended on the weight of the atmosphere to move the pistons which led them to be even heavier. By the nineteenth century, Watt's patent, which had effectively covered all types of steam engines, came to an end and engineers began exploring other designs and were coming to realize that the basic source of power in the steam engine was heat (Cardwell 1995, 209–216). Even based on the older caloric theory of heat engineers, like Sadi Carnot, came to realize that engines would be more efficient if the heat "fell" a greater range, but this required raising the temperature of steam which could be done by increasing its pressure. Early in the century, two inventors, Richard Trevithick in Great Britain, and Oliver Evans in the United States, began building high-pressure steam engines. These had a significant advantage over the low-pressure engines in that they could be much smaller, lighter and had the potential of powering land vehicles. Within a short period of time, a number of inventors, including Trevithick, were experimenting with steam-powered vehicles that moved over rails (Cardwell 1995, chap. 10; Deane 1969, chap. 5). Often seen as the founders of the railways were George Stephenson and his son Robert. George built a number of engines for short-distance collier railways that moved coal from mines. In 1821, George with the help of his son began work on the Stockton to Darlington rail line that connected a collier with the River Tees. The railway opened in 1825, carrying not only coal but also passengers, with engines designed by George and built by Robert's company. It was still not clear if steam-powered engines were superior to a horse-drawn system. In 1829, the directors of the Liverpool and Manchester Railway that was nearing completion decided to conduct the Rainhill Trials to determine what type of engine would be used on the rail line. There were four steam engines, including the *Rocket* designed by Robert, and one horse-powered vehicle. After the *Rocket* won the Rainhill Trials, the railway age began to dramatically and rapidly grow. The railways played an important role in unifying Victorian Britain. Railways could go places where canals could not be built and they were much more efficient than normal roads, or even turnpikes. A major advantage of railways was speed. Even the *Rocket* reached speeds of 30 mph and by mid-century locomotives were reaching more than 70 mph. Before the railway, most people lived and died within about ten miles of where they were born, but after the railways people could travel almost anywhere in a very short amount of time. The railways were also crucial to the development of industrialization. Manufacturing requires the transformation of raw materials into useful products and then selling those products to some market, but it is rare that raw materials, manufacturing, and markets all exist in the same location, so efficient transportation becomes essential for manufacturers.

Finally, one of the most transformative technologies to emerge from the Industrial Revolution was the electric telegraph (Gleick 2011; Cardwell 1995, 250–252; G. Wilson 1859). Before invention of the electric telegraph, the vast majority of messages had to be physically carried from place to place. Throughout history, there had been attempts to speed communication using drums, smoke signals, or light signals. During the French Revolution, Claude Chappe and his three brothers devised an optical telegraph (literally sky writing), using a series of towers with a semaphore-like system at the top. The system did speed communication with Chappe claiming that he could send a signal 475 miles using 120 stations in just twelve minutes, but actual messages took much longer and were often interrupted by bad weather and of course by darkness (Gleick 2011, 129–136). As early as the eighteenth century, there had been suggestions about using static electricity as a type of telegraph but nothing practical ever developed. With the invention of the battery (or pile) that produced a continuous flow of electricity, there were attempts to send a message by using twenty-seven wires to cause electrolysis of water, but the system was too slow and too expensive to be practical. Oersted's discovery of electromagnetism in 1820 immediately led to new interest in an electric telegraph, but there were a number of practical problems that had to be overcome. Early batteries were weak and the electrical resistance in wires meant that signals could only be transmitted very short distances. These problems were solved almost simultaneously by William Cooke and Charles Wheatstone in England and Samuel Morse in the United States. They all benefited from the work of Joseph Henry, one of America's leading scientists, who wrote an article on how to strengthen batteries, strengthen the electric signals by insulating the coiled wires, and creating a "repeater" circuit or a "relay" that would use the last amount of electric current to close a circuit bringing in a new battery (like a Pony Express relay). Morse claims that he came up with the repeating circuit by himself but some doubt this and it seems clear that Cooke and Wheatstone met with Henry when he visited England. In both England and the United States, the development and spread of the telegraph was assisted by the development of railways (the Cooke and Wheatstone system was first demonstrated in 1838 on the Great Western Railway, and Morse's first system was built in 1842 along the Baltimore & Ohio rail line). In order to efficiently manage trains and prevent accidents, railways needed to know where trains were and what parts of the tracks were open. Cooke and Wheatstone's first railway telegraph was almost a direct copy of Oersted's original discovery. It consisted of five wires beneath five compass needles. When a pair of wires were charged, two of the needles would be deflected and point toward a letter of the alphabet on a diamond-shaped board. The use of five wires made the system not very practical but Cooke and Wheatstone made a simpler system with just one wire that would point toward either "line clear" or "line blocked." Morse's system was much simpler in its technology and only required one wire but required the use of code in order to send detailed messages.

Because of its close connection to the railways, the telegraph rapidly expanded as the rail system expanded. But quickly the telegraph came to be seen as something more than a tool to manage rail lines. Newspapers and financial institutions came to adopt the telegraph as did the military. The telegraph standardized time led people to see weather as a global, not local, phenomenon allowed the police to outrun an escaping murder and was used by chess players to play games with players in other cities (Gleick 2011, 144). By 1851, the telegraph network began to become an international network with the completion of an underwater cable between England and France and in 1858 the completion of the Atlantic cable between Great Britain and North America (although the cable only worked for three weeks it set the stage for a more permanent cable in 1866). Like the railways, the telegraph was thought to unify the world. James Gleick quotes the American writer Nathaniel Hawthorne saying that electricity has transformed matter into a "great nerve" and that the "globe is a vast head, a brain" (Gleick 2011, 125). Alfred Smee, a surgeon who also conducted electrical research, compared batteries to brains and telegraph wires to nerves (Gleick 2011, 126).

Institutions

During the first half of the nineteenth century, not only were dramatic conceptual changes taking place in science and technology, but their organizational and the institutional structures were also being transformed (Morrell 1990, 980–989; Basalla, Coleman, and Kargon 1970, 1–21). Ever since its founding in 1660 (it gained a royal charter in 1662), the Royal Society of London for Improving Natural Knowledge had dominated British science. The Royal Society was influenced by Francis Bacon's Salomon's House, which was at the center of the utopian island of Bensalem described in his *New Atlantis*. The Salomon's House functioned as a central research institution for the island with laboratories, towers, deep mines, and museums in which to conduct experimental research into mechanics, optics, acoustics, sources of power, navigation, textiles, mining, metallurgy, agriculture, medicine, and pharmacology, to name but a few areas (Channell 2019, 31–33). The *Philosophical Transactions of the Royal Society*, begun in 1665 under the guidance of Henry Oldenburg, secretary of the Society, served to report many of the most important scientific discoveries, including Newton's *Principia*, and Newton would later serve as president of the Royal Society until his death. By the late eighteenth century, the Royal Society was going into a decline (Morrell and Thackray 1982, chap. 2; Basalla, Coleman, and Kargon 1970, 7–10). The membership was quite limited and by the late eighteenth century, many of the members were no longer scientists but aristocrats making the Royal Society more of a club than a scientific society. Also, by the beginning of the nineteenth century London was no longer the primary center of British science. Places like Manchester and Birmingham in England, and Edinburgh and Glasgow in Scotland were becoming major scientific centers. The nineteenth century also saw the beginning of more specialized scientific organizations, such as the Geological Society of London, founded in 1807, the Astronomical Society of London, founded in 1820, the Zoological Society of London, founded in1826, the Chemical Society of London, founded in 1841, as well as the Pharmaceutical Society, founded the same year. The Royal Society had always seen itself as representing all science and rejected specialized subsections.

When attempts to reform and Royal Society met with only limited success and with the publication in 1830 of Charles Babbage's *Reflections on the Decline of Science in England and on Some of Its Causes*, a group of leading British scientists, including William Harcourt, David Brewster, Charles Babbage, and William Whewell, organized the British Association for the Advancement of Science (B.A.A.S.) in 1831 modeled on the German *Gesellschaft Deutscher Naturforscher un Ärtze* (Society of German Scientists and Physicians) (Morrell and Thackray 1982, chap. 3). Harcourt also was inspired by the works of Francis Bacon. According to Jack Morrell and Arnold Thackray: "Harcourt took the title of the new Association from Bacon's *Advancement of Learning*, its methodology from the *Novum Organum*, and its programme from the *New*

Atlantis" (Morrell and Thackray 1982, 282). Like Salomon's House, the Association would encourage open and collaborative research. Unlike the elitist Royal Society, the British Association aimed to be more open and democratic by holding its meetings in different British cities each year. The Association also established a number of committees that focused on specific sciences which later became more formalized sections. The British Association has been seen by some as the beginning of the professionalization of British science. It was at the 1833 Cambridge meeting of the British Association where William Whewell coined the term "scientist," to refer to those who had previously been called "natural philosophers" (Morrell and Thackray 1982, 20). But, the British Association also would play an important role in publicizing and popularizing science. While section meetings focused on technical issues, the annual presidential address was aimed at the general public and attracted large audiences and became part of a larger movement of public scientific lectures that would become a hallmark of Victorian science (Basalla, Coleman, and Kargon 1970, 4–7).

During the first half of the nineteenth century, scientific and technological education was undergoing important changes in Great Britain (Brock 1990; Birse 1983; Timmons 1983; Cotgrove 1958; Emmerson 1973; Stephens 1972; Horn 1967; Hilken 1967; Cardwell 1957). While today we see a close association between science and universities, this was not always the case especially in eighteenth-century England. The ancient universities of Oxford and Cambridge taught little science or natural philosophy. Their curricula had changed little since their founding. They both focused on the classics and Cambridge did focus on mathematics. It must be remembered that Newton was Lucasian Professor of Mathematics not Natural Philosophy. What little science that was taught was usually some version of Aristotle as part of the seven liberal arts (Brock 1990, 952). Many of the leading Victorian scientists were self-taught or trained in Germany or France. Those who were trained in English universities were often critical of that education (Basalla, Coleman, and Kargon 1970, 11). It was the state of science in the English universities that led Babbage to publish his *Reflections on the Decline of Science in England*. Until 1857, the English universities were controlled by the Church of England and only Anglicans could attend or teach in the universities. Their role was to provide a broad classical education to the upper class and they had little incentive to offer training that might benefit the new emerging merchant and industrial classes. Some new professorships were added in the eighteenth century in areas such as chemistry, astronomy, and biology, but they had little impact on the overall classical focus, although G.B. Airy's courses on experimental physics touched on some engineering subjects and his work influenced many of the leading engineers to the time, such as Robert Stephenson and Isambard Kingdom Brunel (Emmerson 1973, 111–115). Some attempts were made at reforming the universities in 1847 after Prince Albert was named Chancellor of Cambridge but real reform had to wait until the 1850s when Parliament forced reforms based on a Royal Commission Report (Basalla, Coleman, and Kargon 1970, 13). During the 1850s, Oxford and Cambridge began offering degrees in natural science (Brock 1990, 952).

The narrow classical curriculum at Oxford and Cambridge led some dissenters and radicals inspired by Jeremy Bentham, to found London University in 1826 (later named University College, London which then became part of the University of London along with King's College, an Anglican college) (Brock 1990, 952–953; Emmerson 1973, 116–119; Basalla, Coleman, and Kargon 1970, 11). Being sectarian and nonresidential, it was open to a much larger group of students than Oxford and Cambridge. Both science and engineering became important parts of the curriculum at London University. Charles Lyell, the famous geologist, taught geology and Charles Wheatstone, coinventor of the telegraph, taught manufacturing art and machinery (Emmerson 1973, 118–119). When a chair of practical chemistry was created in 1845, Prince Albert raised funds to create the Royal College of Chemistry, headed by Augustus W. Hoffman, a student of Justus Liebig (Basalla, Coleman, and Kargon 1970, 11–12). This helped to introduce German laboratory science into England.

The situation for education in both science and technology was much different in Scotland. Unlike England, Scottish universities were not under the control of the Church of England, and Scotland had a much more democratic outlook concerning education (Davie 1961). There was much more support for science, especially in Edinburgh, where courses in medicine had been taught since the early eighteenth century and by 1746 a medical school had effectively been established at the University (Horn 1967, 41–46). Scotland also had a tradition of supporting technical education. The mechanics' institutes movement had some of its roots in Scotland when George Birkbeck who had recently been named Professor of Chemistry and Natural Philosophy at the Andersonian Institution in Glasgow, allowed workers to attend some of his classes in 1799 (Channell 2019, 61; Emmerson 1973, 100–110). The Andersonian Institution (it would have a number of different names) was founded in 1796 at the bequest of John Anderson, a Professor of Natural Philosophy at Glasgow University and a believer that natural philosophy should be practical and not simply mathematical. Birkbeck developed a special course for workers and artisans. In the first half of the nineteenth century, a number of self-help organizations began to arise that included lectures and libraries aimed at the working class. The Mechanical Institution was formed in London in 1817, and a group at the Andersonian Institution established the Glasgow Mechanics' Institution in 1823. By 1850, there were more than 600 mechanics' institutes throughout Great Britain, and all provided workers with some type of scientific education.

The French had established a number institution of higher education that focused on technology during the eighteenth century, including the famous *École Polytechnique*, but there were no universities in Great Britain that taught engineering or technology until 1840 (Channell 2019, 61–62). A chair of engineering had been established at University College London in 1828 but the holder resigned soon after in a dispute over salary (Buchanan 1985, 221). In 1840, Queen Victoria created a Regius Chair of Civil Engineering and Mechanics at Glasgow University in order to honor James Watt, who had been a mathematical instrument maker at the university when he made his improvements on the steam engine (Channell 2019, 113). The chair was not an immediate success. The first holder of the chair, Lewis D.B. Gordon, met with resistance from the professors of natural philosophy, mathematics, and chemistry, who all saw the chair as infringing on their areas. During some periods, Gordon had trouble filling his classes, and the teaching of engineering did not play a significant role in the University until the appointment of W.J.M. Rankine in 1855. In between Gordon and Rankine, C.B. Vignoles, a well-known engineer, had been appointed to a chair in civil engineering at University College, London in 1841, and in 1846 a chair was established in the mechanical principles of engineering and filled by the famous engineer Eaton Hodgkinson (Emmerson 1973, 118–119). Also, as we will discuss in more detail later, in 1855 Queen Victoria named George Wilson to become Regius Professor of Technology at the University of Edinburgh.

Finally, the organization of British science and technology underwent changes that were associated with the establishment of new science and industrial museums (Kriegel 2007; Yanni, 1999). The development of new museums represented a new location for the practice of science and technology. Museums had arisen from the collections of the wealthy who not only had art decorating their homes but also had collections of other objects, so-called "cabinets of curiosities." These were collections of cultural artifacts, minerals, fossils, stuffed animals, reptiles and fish, and skeletons. While the primary purpose of these collections was display, they also served as research collections. By the nineteenth century, these collections were moving into more public spaces – museums – and with that move the museums became new centers for the production and dissemination of natural knowledge (Yanni 1999, 1). In the middle of the eighteenth century, Hans Sloane's large collection of objects was donated to the British people and became the basis of the British Museum which opened in 1753. By the first half of the nineteenth century, museums focusing on science began to be established in London. In 1813, the Royal College of Surgeons opened the Hunterian Museum which contained a large natural history collection as well as a large medical collection. In 1835, the Museum of Economic Geology (later renamed the Museum of Practical Geology) opened. The Museum was associated with the newly established Geological Survey, the purpose of which was to discover and map Britain's mineral resources. It also was associated with the new School of Mines, and the Mining Record Office. Both the Hunterian Museum and the Museum of Practical Geology differed from today's concept of museums, in that they were not primarily designed for display and entertainment but were seen as types of research laboratories. In fact, the Hunterian was not open to the public without a letter of reference from a member of the College of Physicians, and the Museum of Practical Geology basically served as a research collection for the Geological Survey and the School of Mines.

During the middle of the nineteenth century, a number of new museums began to be established throughout Great Britain. In 1845, an extension of the Museum of Economic Geology was established in Dublin (it was later renamed the Museum of Irish Industry). A major stimulus to the building of new museums was the Great Exhibition of the Industry of All Nations held in the Crystal Palace in Hyde Park in 1851. As a result of the Great Exhibition, the Board of Trade of the British government established the Department of Science and Art in 1853. Drawing on some of the profits generated by the Great Exhibition money from parliament and some of the objects exhibited in the Crystal Palace, the Department began a widespread program to promote education in science and technology in Britain and Ireland. Establishing new museums became a major part of the Department's program. It took control of the Museum of Irish Industry and established the South Kensington Museum that would become the forerunner of the Victoria and Albert Museum, and the Science Museum. More important for our story, the Department of Science and Art obtained funding to establish the Industrial Museum of Scotland in 1855 and named George Wilson as its first director.

As we have seen, science and technology were undergoing significant changes during the first half of the nineteenth century. Chemistry was attempting to explain new chemical elements and processes in terms of chemical atoms. Natural history was trying to explain a wide range of geological phenomena and new fossil evidence according to some "unity of plan." The focus on the science of energy in natural philosophy provided new unifying laws of thermodynamics to explain not just heat but all natural phenomena. The Industrial Revolution was literally unifying the world through railways and the telegraph. As we will see, this belief in a unity of nature was closely tied to ideas associated with natural theology and a theology of nature that were particularly influential during the early Victorian period. As we will see, George Wilson, whose guiding motto was "Unity in Variety," participated in and was shaped by all of these changes and a study of his life will hopefully bring about new insights to science and technology in the early Victorian age.

References

Basalla, George, William Coleman, and Robert Kargon, editors 1970. "Introduction." In *Victorian Science*, 1–21. Garden City, NY: Anchor Books.

Beer, John J. 1959. *The Emergence of the German Chemical Dye Industry.* Urbana: University of Illinois Press.

Birse, Ronald M. 1983. *Engineering Education at Edinburgh University, 1673–1883.* Edinburgh: School of Engineering, University of Edinburgh.

Bowler, Peter J. 2009. *Evolution: The History of an Idea*, rev. ed. Berkeley: University of California Press.

Brock, W.H. 1990. "Science in Education." In *Companion to the History of Modern Science*, edited by Robert C. Olby, Geoffrey N. Cantor, John R.R. Christie, and Michael J.S. Hodge, 946–959. London: Routledge.

Brose, Eric Dorn. 1998. *Technology and Science in the Industrializing Nations, 1500–1914.* Atlantic Highlands, NJ: Humanities Press.

Buchanan, R.A. 1985. "The Rise of Scientific Engineering in Britain." *British Journal for the History of Science* 18: 218–233.

Cardwell, Donald S.L. 1957. *The Organisation of Science in England.* London: William Heinemann.

Cardwell, Donald S.L. 1971. *From Watt to Clausius: The Rise of Thermodynamics in the Early Industrial Age.* Ithaca, NY: Cornell University Press.

Cardwell, Donald S.L. 1995. *The Norton History of Technology.* New York: W.W. Norton.

Channell, David F. 1991. *The Vital Machine: A Study of Technology and Organic Life.* New York: Oxford University Press.

Channell, David F. 2017. *A History of Technoscience: Erasing the Boundaries between Science and Technology.* London: Routledge.

Channell, David F. 2019. *The Rise of Engineering Science: How Technology Became Scientific.* Dordrecht: Springer.

Clow, Archibald, and Nan L. Clow. 1952. *The Chemical Revolution: A Contribution to Social Technology.* London: Batchworth Press.

Cotgrove, Stephen. 1958. *Technical Education and Social Change.* London: George Allen & Unwin.

Davie, George Elder. 1961. *The Democratic Intellect: Scotland and Her Universities in the Nineteenth Century.* Edinburgh: Edinburgh University Press.

Deane, Phyllis. 1969. *The First Industrial Revolution.* Cambridge: Cambridge University Press.

Donovan, Arthur, ed. 1988. "The Chemical Revolution: Essays in Reinterpretation." *Osiris* 4: 4–12.

Doran, B.G. 1975. "Origins and Consolidation of Field Theory in Nineteenth-Century Britain: From the Mechanical to the Electrodynamic View of Nature." *Historical Studies in the Physical Sciences* 6: 133–260.

Eddy, Daniel, Seymour Mauskopf, and William R. Newman. eds. 2014. "An Introduction to Chemical Knowledge in the Early Modern World." *Osiris* 29: 1–15.

Emmerson, George S. 1973. *Engineering Education: A Social History.* Newton Abbot: David & Charles.

Gillispie, Charles Coulston. 1960. *The Edge of Objectivity: An Essay in the History of Scientific Ideas.* Princeton, NJ: Princeton University Press.

Gleick, James. 2011. *The Information: A History, a Flood.* New York: Pantheon Books.

Guerlac, Henry. 1975. *Antoine-Laurent Lavoisier: Chemist and Revolutionary.* New York: Charles Scribner's Sons.

Harman, Peter M. 1982. *Energy, Force and Matter: The Conceptual Development of Nineteenth-Century Physics.* Cambridge: Cambridge University Press.

Hesse, Mary B. 1965. *Forces and Fields: A Study of Action at a Distance in the History of Physics.* Totowa, NJ: Littlefield, Adams.

Hilken, Thomas John Norman. 1967. *Engineering at Cambridge University, 1783–1965.* Cambridge: Cambridge University Press.

Holmes, Richard. 2014. "In Retrospect: On the Connexion of the Physical Sciences." *Nature* 514: 432–433.

Horn, David Bayne. 1967. *A Short History of the University of Edinburgh, 1556/1889.* Edinburgh: Edinburgh University Press.

Ihde, Aaron. 1964. *The Development of Modern Chemistry.* New York: Harper & Row.

Jenkins, Bill. 2016. "Neptunism and Transformation: Robert Jameson and other Evolutionary Theories in Early Nineteenth-Century Scotland." *Journal of the History of Biology* 49: 527–557.

Kriegel, Lara. 2007. *Grand Designs: Labor, Empire, and the Museums of Victorian Culture.* Durham, NC: Duke University Press.

Kuhn, Thomas. 1959. "Energy Conservation as an Example of Simultaneous Discovery." In *Critical Problems in the History of Science*, edited by Marshall Claggett, 321–356. Madison: University of Wisconsin Press.

Kuhn, Thomas. 1977. *Essential Tension.* Chicago: University of Chicago Press.

Landes, David S. 1969. *The Unbound Prometheus: Technological Change and Industrial Development in Western Europe from 1750 to the Present.* Cambridge: Cambridge University Press.

Laudan, Rachel. 1990. "The History of Geology, 1780–1840." In *Companion to the History of Modern Science,* edited by Robert C. Olby, Geoffrey N. Cantor, John R.R. Christie, and Michael J.S. Hodge, 314–325. London: Routledge.

Lovejoy, Arthur O. 1936. *The Great Chain of Being: The History of an Idea.* New York: Harper.

MacLeod, Christine. 1988. *Inventing the Industrial Revolution: The English Patent System, 1660–1800.* Cambridge: Cambridge University Press.

Miller, David Phillip. 2004. *Discovering Water: James Watt, Henry Cavendish and the Nineteenth-Century 'Water Controversy'.* London: Routledge.

Morrell, Jack .B. and Arnold Thackray. 1982. *Gentlemen of Science: Early Years of the British Association for the Advancement of Science.* Oxford: Oxford University Press.

Morrell, Jack .B. 1990. "Professionalization." In *Companion to the History of Modern Science,* edited by Robert C. Olby, Geoffrey N. Cantor, John R.R. Christie, and Michael J.S. Hodge, 980–989. London: Routledge.

Mokyr, Joel. 2009. "Intellectual Property Rights, the Industrial Revolution, and the Beginnings of Modern Economic Growth." *The American Economic Review* 9: 349–355.

Olson, Richard. 1975. *Scottish Philosophy and British Physics, 1750–1880: A Study in the Foundations of the Victorian Scientific Style.* Princeton, NJ: Princeton University Press.

Perrin, Carleton. 1990. "The Chemical Revolution." In *Companion to the History of Modern Science,* edited by Robert C. Olby, Geoffrey N. Cantor, John R.R. Christie, and Michael J.S. Hodge, 264–277. London: Routledge.

Rehbock, Phillip F. 1983. *The Philosophical Naturalists: Themes in Early Nineteenth-Century British Biology.* Madison: University of Wisconsin Press.

Schofield, Robert E. 1970. *Mechanism and Materialism: British Natural Philosophy in an Age of Reason.* Princeton, NJ: Princeton University Press.

Secord, James. 1991. "Edinburgh Lamarckians: Robert Jameson and Robert Grant." *Journal of the History of Biology* 24: 1–18.

Sloan, Phillip R. 1990. "Natural History, 1670–1802." In *Companion to the History of Modern Science,* edited by Robert C. Olby, Geoffrey .N. Cantor, John R.R. Christie, and Michael J.S. Hodge, 295–313. London: Routledge.

Smith, Crosbie and M. Norton. Wise. 1989. *Energy and Empire: A Biographical Study of Lord Kelvin.* Cambridge: Cambridge University Press.

Smith, Crosbie. 1998. *The Science of Energy: A Cultural History of Energy Physics in Victorian Britain.* Chicago: University of Chicago Press.

Somerville, Mary. 1849. *On the Connexion of the Physical Sciences*, 8th ed. London: John Murray.

Stephens, Michael. 1972. "British Artisan, Scientific and Technical Education in the Early 19th Century." *Annals of Science* 29: 87–98.

Timmons. George. 1983. "Education and Technology in the Industrial Revolution." *History of Technology* 8: 135–149.

Toynbee, Arnold. 1887. *Lectures on the Industrial Revolution of the 18th Century in England*, 2nd ed. London: Rivingtons.

Williams, L. Pearce. 1971. *Michael Faraday: A Biography.* New York: Simon and Schuster.

Wilson, George. 1859. *Electricity and the Electric Telegraph.* London: Longman, Brown, Green, Longman and Roberts.

Wilson, Jessie Aitken. 1860. *A Memoir of George Wilson.* Edinburgh: Edmonston and Douglas.

Yanni, Carla. 1999. *Nature's Museums: Victorian Science and the Architecture of Display.* Baltimore, MD: Johns Hopkins University Press.

Zukerfeld, Mariano. 2014. "On the Link between the English Patent System and the Industrial Revolution: Economic, Legal and Sociological Issues." *Intersect* 8: 1024.

2 From Medicine to Chemistry

Wilson's Early Years

Influence of Family

On February 21, 1818, George Wilson and his twin brother John were born in Edinburgh to Archibald Wilson, a wine merchant, and Janet Aitken Wilson, the daughter of a land surveyor from Greenock (J. Wilson 1866). George and his twin brother were the middle children of a family that eventually totaled eleven children, although five died before the age of five. Near the end of his life, he wrote: "I saw in my early childhood or boyhood, so many little brothers and sisters die, that the darkness of those scenes, and the anguish of father and mother, made an indelible impression on me" (J. Wilson 1866, 4). But his sister Jessie notes that his early childhood was open to as much joy as sorrow. His mother especially valued learning and took charge of her children's education from an early age. George began elementary school at age four, and at age nine he entered the High School of Edinburgh (sometimes called the Royal High School of Edinburgh) which was well known for providing a solid education in the classics, including Latin and Greek. This early education shows in Wilson's later popular books and lectures that almost always contain some literary reference, especially to John Milton and William Shakespeare. According to his sister, from a very early age George exhibited an interest in natural history and also developed an early interest in writing poems. When an aunt died and left four cousins as orphans, the Wilson family took them in. The household contained a large number of pets, including a hedgehog, a tortoise, and a bull terrier. George's older brother Daniel, who would go to become one of the leading archeologists in Scotland, then one of the leading anthropologists in North America, and eventually the President of the University of Toronto, shared interest with George in natural history. Their sister writes:

> George's brother Daniel recalls many excursions, on Saturdays and other holidays, to places of interest in the neighbourhood of Edinburgh. Both George and his brothers were good pedestrians, and many a happy day they spent in visiting picturesque ruins, bring home botanical of geological specimens, as tangible tokens of a day's pursuits.
>
> (J. Wilson 1866, 9)

DOI: 10.4324/9781003212218-2

Aside from an interest in collecting natural objects, they became interested in the science behind these objects and the brothers and some friends formed a "Juvenile Society for the Advancement of Knowledge," overseen by their mother. At their weekly meeting, the members would read papers on natural history, mechanics, astronomy, and other scientific topics.

Influence of the University

At age fourteen, George left the High School to enter the Medical School of the University of Edinburgh and also began a four-year apprenticeship in the laboratory of the Royal Infirmary. In an address to students later in life, Wilson recalled some of his first experiences in the Infirmary. He said: "When a young student first visits the hospital, his faith in God as the wise and merciful designer of man's body, must, in sympathizing natures, undergo a painful shock" (J. Wilson 1866, 12). He then went on to describe patients awaiting amputations, suffering from cancer, or heart disease, or lupus. He then goes on to describe observing his first surgical operation which was the amputation of a sailor's leg and notes: "The spectacle, for which I was quite unprepared, sufficiently horrified a boy fresh from school, especially as the patient underwent the operation without the assistance of anaesthetics, which were not introduced into surgical practice till many years later" (J. Wilson 1866, 14). These first experiences probably helped to dissuade him from following a career in clinical medicine.

Wilson's experience in the University was entirely different from that of the Infirmary. The Medical School did not simply train students in the narrow techniques of medical practice, but provided them a broad education in the sciences. At the time he entered the University, there was a broad debate taking place in Great Britain concerning the proper goals of a university education and this debate would help to shape Wilson's education. As outlined by George Davie, the debate was between the English idea of education and the Scottish version (Davie 1961). The Act of Union in 1707 led to Scotland giving up much of her political and economic independence to England but retaining more independence in religious, legal, and educational matters. This led some to argue that the educational independence of the Scottish universities was threatening to a unified "British way of life" (Davie 1961, 3). Others saw the independent Scottish universities as a way to resist a cultural takeover by the English. The debate began in 1830 when a Royal University Commission recommended reforming the Scottish universities by raising the entrance age and focusing more on specialized education rather than general education which would make the Scottish universities closer to the English model. The major champion of the Scottish educational ideal was George Jardine, Professor of Logic and a student of Adam Smith at Glasgow University, whose *Outlines of a Philosophical Education* (1825) set forward the differences between the Scottish and English systems of education (Davie 1961, 9–20). Several of Jardine's students became friends of George Wilson. In England, students had to choose between a specialized set of courses in the classics at Oxford or a set of mostly mathematical courses at Cambridge. The primary difference of Jardine's plan was that it took a more broad and philosophical approach to education. Much of his philosophy was based on Common Sense philosophy, German Idealism, or a combination of the two as seen in the work of William Hamilton.

During the second half of the eighteenth century and the first half of the nineteenth century, Common Sense philosophy came to strongly influence Scottish education as well as intellectual life in general (Olson 1975). One of the leading founders of Common Sense philosophy was Thomas Reid, who was Professor of Philosophy at Kings College, Aberdeen and a Founder of the Philosophical Society of Aberdeen (Olson 1975, 27). Reid's ideas would be later developed and revised in the nineteenth century by Dugald Stewart, who was Professor of Moral Philosophy at the University of Edinburgh. Reid's ideas arose as a reaction to the skeptical philosophy of David Hume, which many saw as supporting atheism, and the works of David Hartley and Joseph Priestley which were seen as supporting materialism and, therefore, atheism (Olson 1975, 28–35). Hume's idea that causality could never be empirically proven simply using our senses raised questions concerning a person's ability to understand the existence of the external world, while Hartley's and Priestley's materialistic doctrine of the human mind raised questions concerning a person's free will. As an answer, especially to the critique of causality raised by Hume, Common Sense philosophers followed a path similar to Immanuel Kant's "Copernican Revolution in Philosophy" that he used to address Hume's critique of causality (Olson 1975, 14–16). Rather than assuming that the order of the physical world was the source of knowledge in the human mind, Common Sense argued that our source of knowledge of the natural world is dependent upon the human mind and before we can understand the external world, we must first understand the nature of the human mind. That is, Reid and his followers argued that along with a science of nature, there also must be a science or philosophy, of the human mind. An interesting characteristic of the Common Sense notion of mind was that it was not simply passive, like John Locke's *tabula rasa*, but had intuitive properties, Principles of Common Sense or First Principles (what Kant labeled categories of the mind), that helped interpret the information concerning the external world that the mind was receiving from the senses (Olson 1975, 29–30). These principles, such as the idea of causality, and the notions of space and time, could not be empirically proven but were rather intuitive and helped to give meaning to empirical information.

A fundamental characteristic of the Common Sense philosophy was its strict adherence to the empirical method as a way to guard against errors of reasoning that might arise from the use of hypotheses that could not be confirmed by the senses or experiments (Olson 1975, 34–35). As a result, Common Sense philosophy relied heavily on the works of Francis Bacon and Isaac Newton. Bacon's inductive method provided a way to build a theory without the reliance on hypotheses (later Dugald Stewart would be more open to a limited role of hypotheses in the formulation of experiments), and Newton's refusal to "feign" a hypothesis concerning the nature of gravity became a model for developing a theory without hypotheses (Common Sense philosophers would reject Newton's use of the hypothesis of an aether that he put forward in the *Opticks* to explain gravity). By the mid-nineteenth century, Sir William Hamilton, who was a colleague of Wilson's, attempted to bring together Common Sense philosophy and Kantian philosophy.

Students in Scottish universities were given a grounding in philosophy that was strongly influenced by Common Sense. For example, Jardine argued that much of philosophy should be concerned with the Theory of Knowledge, especially perception, universals, and causality (Davie 1961, 11). This led many of the philosophy classes to focus on psychological problems. This clearly reflected the influence of Common Sense philosophy. For example, in a course taught at Edinburgh by John Wilson (aka Christopher North) in the 1820s, one of the problems given to students was "The Senses of Sight and Touch" (Davie 1961, 18). As we will see later, Wilson often referred to the hand and the eye as being symbols of industrial science and he would write a popular book, *The Five Gateways of Knowledge*, that focused on the five senses.

This philosophical orientation found its way outside of the arts curriculum and into the science and medical courses. Davie quotes L.J. Saunders as saying that during the period from 1815 to 1840 when subjects such as chemistry, geology, and physiology were developing: "The critical approach was indeed marked in the Scottish intellectual inheritance from the eighteenth century enlightenment, and the philosophical tradition was now spreading out into exposition and research in the physical and natural sciences" (Davie 1961, 21). In addition to this philosophical tradition, another distinguishing characteristic of Scottish university education compared with the English system was that a large amount of that education took place outside of the formal lectures, in tutorials and student societies, such as debating clubs (Davie 1961, 14–16). Although Wilson, as a medical student, did not take the formal philosophy courses offered in the arts curriculum, many of his professors in the natural sciences had taken those courses as students and that philosophical bias may have likely entered their own teaching. Davie notes: "There was in the Faculty of Medicine, too, an analogous tendency to keep alive questions of first principles as a guide to research and practice" (Davie 1961, 23). He goes on to note that in Scotland, there was no conflict between the more traditional studies and the newer specialized fields such as chemistry and physiology or sociology and history because: "the legal and medical studies had already – in the persons of the legendary father-figures – accepted the distinctive role in Scotland of the early philosophical training, and tried to turn it to advantage in the specialized disciplines" (Davie 1961, 24). In addition, as we will see, as a student Wilson was very involved in debating societies and scholarly clubs.

The debate over the English versus the Scottish approach to university education also played out within the University of Edinburgh beginning in the

1830s between J.D. Forbes, Professor of Natural Philosophy on one side, and David Brewster, a Scottish physicist who would go on to become Principal of St. Andrews University and then Principal of the University of Edinburgh, and William Hamilton, the Scottish philosopher and Professor of Civil History (Davie 1961, chap. 8). While all parties involved agreed on the value of Scotland's focus on general education, they disagreed as to how this general education should apply to the new world that was developing. Forbes felt that the focus on philosophy provided a hindrance to the development of a more mathematical approach to physics as had been done at Cambridge. On the other hand, Brewster argued that mathematical physics was not the model for all of the sciences and that the role of mathematics in the sciences should be reduced in order to develop sciences that differed from Cambridge. In testimony to the Royal University Commission during the late 1820s Brewster said:

> The general idea which has occurred to me, in regard to Natural Philosophy, is that, in place of a course of lectures such as has been delivered for the past thirty years, a course perfectly popular and experimental ought to be substituted. An ordinary student cannot derive any satisfaction from the present system, owing to the mixture of experimental and mathematical instruction in the class. In the lectures which are delivered, there occur many mathematical demonstrations which no person, but one or two, can follow.
>
> (Davie 1961, 171–173)

Brewster saw a positive outcome to reducing the mathematization of science. With less dependence on higher mathematics, the teaching of science could focus on the experimental and practical sides of science. He noted that this approach had been successful in the "working-men's University" in Glasgow, probably referring to Anderson's University (also called Anderson's Institution or the Andersonian) and the Glasgow Mechanics' Institution (Davie 1961, 172). According to Davie:

> In short, Brewster apparently had in mind a reform which would humanise scientific instruction, in the sense of emphasising its relevance to practical life, and which, in that way, would have been quite in line with the educational traditions of Scottish democracy.
>
> (Davie 1961, 172)

Brewster's fear of the Cambridge higher mathematical approach was that science would lose its connection to the practical arts which had played an important role in Scotland.

Brewster did not get to put his reforms into practice, at least for several years. The Chair of Natural Philosophy went to Forbes, and Brewster became Principal of St. Andrews University. Ironically, many years later, in 1858, Forbes would offer a compromise between the Cambridge and Edinburgh approach which he argued could be accomplished by a union between a mathematical and an experimental approach to science (Davie 1961, 179–190). In the end, neither the English nor the Scots supported Forbes's compromise and in a further irony, Brewster would return in 1859 as Principal of the University of Edinburgh. In any case, Forbes's move toward mathematization of the sciences had its major impact on natural philosophy and much less, if any, impact on the sciences associated with medicine, such as anatomy, chemistry, natural history, and botany, which tended to continue to focus on both the philosophical and the practical.

Influence of the Medical School Faculty

The best understanding of Wilson's education at the University of Edinburgh can be gained through his *Memoir of Edward Forbes* (G. Wilson and Geikie, 1861). Forbes entered the University of Edinburgh just a year before Wilson, with the intention to study medicine. For the rest of their lives, the two would be close friends and Forbes would go on to become Professor of Botany at King's College in London, paleontologist with the Geological Survey, and eventually Regius Professor of Natural History at the University of Edinburgh. Soon after beginning his position in Edinburgh, Forbes tragically died, and Wilson agreed to write his *Memoir*. Wilson himself would die not too long after but was able to write the first part of the *Memoir* that included details of Forbes's student years. Since they were only a year apart and taking many of the same medical school classes, Wilson's descriptions of courses give us insights into Wilson's own education at Edinburgh.

During Wilson's time as a medical student, the basic science courses were undergoing some changes. Anatomy especially was being transformed. In his first year, along with courses in natural philosophy and mathematics, Wilson took an anatomy class with John Lizars who gave extra-mural courses associated with the Royal College of Surgeons and the Royal Infirmary (J. Wilson 1866, 17–20). It was common for many of the medical classes to be offered by extra-mural lectures outside the University, and Wilson himself would later offer a number of these courses. Lizars had previously taught Charles Darwin when he attended Medical School at the University from 1825 to 1827. Anatomy was still limited by the inability to legally obtain cadavers for dissection (G. Wilson and Geikie 1861, 92–98). Only the bodies of executed prisoners, suicide victims, and orphans could legally be used for dissections. This led to what was called resurrectionism, or what we would call body snatching. Wilson notes that one of his experiences during this time was seeing a fellow student in the Infirmary with a gunshot wound received while trying to obtain a body (G. Wilson and Geikie 1861, 92). Wilson also notes that the anatomical lecture rooms often had sliding panels or a lift to a garret where bodies could be quickly hidden if a law officer approached.

Often times illegally obtained bodies from graveyards were not enough to meet the demand which led to the notorious case of Burke and Hare in

1828 (MacGregor 1884). William Burke and William Hare ran a boarding house in Edinburgh and when a boarder died of natural causes, they decided to sell his body to Robert Knox, one of the most famous anatomists at the time. In 1828, another boarder became ill and they decided to suffocate him and sell his body to Knox. With no other ill lodgers, the men began to lure poor people back to their lodge and suffocate them, murdering at least sixteen people during 1828. Burke, Hare, and Knox were tried in December of 1828, but only Burke was found guilty after Hare was granted immunity to testify against Burke. Burke was subsequently hanged and ironically his body was used for dissection. After another set of similar murders took place in London, Parliament passed the Anatomy Bill in 1832 which expanded and regulated the use of bodies for dissection. Before the Bill was passed, it fell to Wilson to defend the Bill at one of the University's debating societies. But even with the passage of the Bill, for a time, it was still difficult for medical students to study the human body, and Wilson notes that they often had to rely on skeletal bones passed down from professors to students or to search churchyards for relics of bone. Given that almost no one was in possession of a complete skeleton, students often had to rely on casts or engravings.

During his second year, Wilson took another anatomy class with Lizars along with a class within the University with Alexander Munro (tertius), third in a line of famous anatomists and also a teacher of Darwin. Most importantly, Wilson began his study of chemistry with Thomas Charles Hope who had followed Joseph Black as Professor of Chemistry. Hope had been one of the first chemists in Great Britain to reject the phlogiston theory and accept Antoine Lavoisier's theory of combustion. Phlogiston theory argued that during combustion a substance labeled phlogiston escapes, but Lavoisier argued that combustion was the result of the chemical combination of a material with oxygen. After beginning to teach Lavoisier's new theory while still at Glasgow University Hope spent a summer in Paris and met with Lavoisier. It seems that Hope instilled in Wilson a new interest in experiments and provided a model for the showmanship that Wilson would use later in life (G. Wilson and Geikie 1861, 98–102). Hope was one of the most popular professors of chemistry in Great Britain. In 1823, he taught 559 students, and throughout his career he taught 15,000 students. To keep the attention of such large classes, Hope relied on demonstrating the principles of chemistry through a range of experiments rather than simply lecturing on the subject. While others had used experimental demonstrations in their classes, Wilson notes that "no one before him, in this country at least, had ventured to give a series of strictly scientific lectures, extending for five days weekly over nearly six months, and each illustrated to the full by experiments" (G. Wilson and Geikie 1861, 99). Hope's ability to popularize scientific ideas may have served as a foundation for Wilson's later popular lectures as Director of the Industrial Museum of Scotland. Also, Wilson notes that every academic area that could be taught through public experimental illustration used such an approach, and the traditional lecture simply became a commentary on the experiment. This came to distinguish Scottish education from the English model where professors would simply "read a lecture." It also probably fits very well with the values of Common Sense philosophy which valued knowledge gained through the senses. While Wilson praised Hope's use of experimental demonstrations in the lecture hall, he also expressed the desire that students should themselves be able to repeat and conduct experiments, but Hope was not in favor of this idea (G. Wilson and Geikie 1861, 100). Wilson notes that some of the wealthier students were able to organize into small societies and find spaces where they could conduct experiments on their own, but there is no evidence that he was part of these groups. There were some extra-mural classes in "practical chemistry" which allowed students to conduct their own experiments. One was offered by David B. Reid, but it is not clear if Wilson took that class, but he would take a practical chemistry course with Kenneth Kemp during his final year.

During the 1834–1835 session, Wilson not only continued his studies of anatomy with Lizars but also began a course in Materia Medica taught by Robert Christison, who would become influential in Wilson's career as a chemist. Christison, who had studied at the University of Edinburgh, became Professor of Medical Jurisprudence in 1822 (G. Wilson and Geikie 1861, 125). In that post, he developed an expertise in the area of toxicology, experimenting with some poisons on himself. He also testified in the famous Burke and Hare murder case which we have already discussed. Later, he would become physician to Queen Victoria which is somewhat ironic since he was vehemently opposed to allowing women into medical schools. In 1832, just as Wilson was entering the University, Christison was appointed Professor of Materia Medica. The field combined two of Wilson's interests, that of chemistry and botany. What also impressed Wilson and may have influenced his future ideas about the Industrial Museum of Scotland was that Christison established a museum that contained botanical, chemical, and mineral specimens (G. Wilson and Geikie 1861, 126). More importantly, there was a laboratory attached to the museum where active chemical research was conducted, especially on the relationship between plants and medicines.

During the 1835–1836 session, Wilson took courses in the practice of medicine and clinical surgery along with more anatomy with Lizars. In May, Wilson not only studied chemistry with Christison but also began the study of botany under Robert Graham. Wilson notes that botany had always been taught as a practical subject at Edinburgh, and Graham continued this tradition (G. Wilson and Geikie 1861, 102–105). This practicality was demonstrated in two ways. Graham made ample use of the Botanic (or "Physic") Garden at the University which was one of the best in Great Britain and many of his lectures took place in the Garden. Another attraction of his course in botany was that in the summer, he would lead students on field trips to Scotland, England, Wales, and Ireland so that students gained knowledge of the flora of the entire United Kingdom. Besides medical students, Graham's course also attracted theology students, and Wilson notes that "in truth, there was a greater mingling of students of all the Faculties, as well as amateurs, at the Botanic Garden, at lecture time, than in any other place connected with the University" (G. Wilson and Geikie 1861, 104). The benefit of demonstrating scientific ideas with concrete examples taken from nature would influence Wilson's later work as Director of the Industrial Museum.

During his final term at the University in 1836–1837, Wilson continued courses in clinical medicine, practical chemistry, with Kenneth Kemp, more botany with Graham, but most importantly he began a study of natural history with Robert Jameson. Jameson's ideas concerning natural history would influence Wilson in two ways: first, he gained direct knowledge from Jameson, and second, he was most likely influenced by Jameson's ideas through his close friend Edward Forbes, who was one of Jameson's leading disciples and would eventually succeed him as Regius Professor of Natural History at the University (Rehbock 1983). On a very practical level, Jameson provided Wilson with examples of both how to, and how not to organize a museum. While the natural history collection predated Jameson, he turned the Natural History Museum into one of the leading museums in Great Britain, and Wilson noted that he had often visited the Museum as a boy (G. Wilson and Geikie 1861, 108–115). The Museum contained a large collection of stuffed animals and birds along with shells, fishes, and reptiles. It was especially strong in mineralogy, which was Jameson's specialty. And while Wilson notes that the Museum was highly instructive and Jameson made good use of it to illustrate his lectures, he did see several deficiencies that would influence how he would later organize his Industrial Museum.

Wilson's first criticism of the Museum was that it was mostly seen as an adjunct to the lecture hall. While it was open to the public, the fee deterred many from visiting it. When inside, visitors who were nonspecialists in natural history would find it difficult to self-interpret because of the organization of the collections and the lack of explanatory labels. Wilson was also critical of the fact that in the mineralogical collection, there was "no series of illustrations representing the stratigraphical, paleontological, or industrial relations of the several geological formations" (G. Wilson and Geikie 1861, 111). He also noted the lack of any drawings, diagrams, or models that would allow mining engineers to learn where to look for various ores and to recognize them once they found them. Wilson also criticized the mineralogy displays for only displaying the minerals in their natural states and not showing their connections to chemistry or the role of minerals in industry. The animals were almost totally displayed as stuffed skins with no description of the underlying anatomy. Skeletons, dissected organs, and microscopic sections were totally missing. Wilson would try to remedy deficiencies like these when he began to organize his Industrial Museum.

While Jameson's Natural History Museum had mostly a negative impact on Wilson's thinking about museums, Jameson's ideas concerning the natural world would have an important effect on Wilson's future thinking, possibly as much through his friendship with Edward Forbes as directly through

Jameson. Wilson says that Jameson "became, what many of his brilliant colleagues failed to become, the founder of a School" (G. Wilson and Geikie 1861, 109). Wilson praises Jameson's knowledge and sympathy for the progress that was taking place in every branch of science. Jameson founded the Wernerian Society at the University which served as a center for discussion for every type of naturalist to discuss the developments in all the sciences. He also served as one of Darwin's teachers. Finally, Jameson served as the editor for over fifty years of the *Edinburgh Philosophical Journal*, and later the *Edinburgh New Philosophical Journal* which was one of the leading international journals that discussed issues in natural history and geology and one in which Wilson would publish almost a dozen articles.

Jameson's most important impact on Wilson was to provide a way to think about the connections between geology and life. As discussed in the first chapter, during the early nineteenth century, geologists were divided into two camps (Jenkins 2016a, 532). One group believed that the earth had not undergone any changes or at least that those changes were cyclical and were sometimes labeled uniformitarians and were represented by James Hutton and later Charles Lyell. The other camp believed that the earth had undergone some type of progressive change. This group was represented by the German geologist Abraham Werner who believed that a slow receding ocean was responsible for the earth's geology, and a little later by the French geologist Georges Cuvier who believed that local floods shaped the earth. Since those advocating some type of change argued for the role of water in that change, they became known as Neptunists. By the end of the eighteenth century, Jameson had come to support the views of Werner and in 1800 he spent a year studying with Werner at the Mining Academy in Freiberg, Saxony. Werner taught that the earth had once been covered by water, but that water was continually receding, possibly because of evaporation (Jenkins 2016a, 532–533). As noted in the first chapter, the first rocks, such as granite, began to chemically precipitate out of the water and were deposited on the ocean's floor. As land masses began to appear because of the receding waters, other rocks such as limestone began to precipitate out, but other rocks, such as sandstone, began to be deposited because of erosion. Therefore, the different chemical compositions of rocks could be explained by the receding waters. According to Bill Jenkins, in addition to Werner's explanation of the role of water, Jameson may have also taught his students that a decline in global temperature may have also played a role in the earth's geology (Jenkins 2016a, 533–534). This idea is similar to that put forward by the French geologist, George Louis Leclerc, Comte de Buffon who argued that the earth began in a molten state but then gradually cooled. In his lectures for 1830, Jameson seems to make an important connection concerning the role played by geological conditions and the distribution of life on the earth. In his lecture notes, he says that in the past "the climate was very different from what it is at present and that at the time Britain was calculated to produce plants and animals requiring a much more considerable temperature than the Island possesses at present" (Jenkins 2016a, 534).

The connection between geology and life would provide Wilson with another important idea that he would use throughout his career – the idea that life develops progressively or even through transmutations. As Jenkins argues, during the early nineteenth century people came to more widely accept that the fossil record seemed to present evidence that life moved from the very simple forms to much more complex forms (Jenkins 2016a, 534–538). In his book, *Elements of Geognosy* (1808), Jameson writes:

> As the water diminished, it appears to have become gradually more fitted for the support of animals and vegetables, as we find them increasing in number, variety and perfection, and approaching more to the nature of this in the present seas, the lower level of the outgoings of strata, or, what is the same thing, the lower level of the water. The same gradual increase of organic beings appears to have taken place on the dry land.
>
> (Jenkins 2016a, 535)

In a set of lecture notes written sometime after 1826, Jameson says:

> Indeed there appears to be a regular & consistent distribution of organic beings through the of this class [Floetz rocks] from the very low species of the earliest strata to the more perfect animals of the newest strata, immediately adjoining the alluvial formation.
>
> (Jenkins 2016a, 535)

In making these statements, Jameson seems to be directly connecting Werner's idea of slow gradual geological change with the subsequent progressive development of life.

While Jameson connected the progressive nature of geology with the progressive nature of the fossil record, Bill Jenkins and James Secord, in two important articles, make the argument that Jameson took the next step to a transformist theory of life (Jenkins 2016a, 538–545; Secord 1991, 1–18). Much of the evidence for this interpretation comes from an article entitled "Observations on the Nature and Importance of Geology," which was published anonymously in the *Edinburgh New Philosophical Journal* in 1826. For at least two decades, it was thought that the article was written by Robert E. Grant, a protégé of Jameson and a supporter of Lamarckian ideas. But in 1991, James Secord, through an analysis of the aims of the article, the context of the article as a whole, and textual parallels, made a convincing argument that Jameson, the editor of the *Journal*, was the actual author of the article (Secord 1991, 4). More recently, Bill Jenkins, in an article on "Neptunism and Transformism," provides some further evidence of Jameson's authorship of the article (Jenkins 2016a, 538–545). While there is no totally definitive proof that Jameson wrote the article, Jenkins writes that even if he did not write the article "there is significant evidence from other sources that he was sympathetic to a transformist interpretation of the history of life" (Jenkins

2106a, 539). It could also be noted that at the time of publication, Jameson was the sole editor of the *Journal*, indicating that he was certainly sympathetic to its argument.

The "Observation" article gives significant support to the ideas of Jean Baptiste Lamarck, the leading French natural historian, and connects his ideas concerning the transmutation of life to Wernerian geology (Jenkins 2016a, 538–539). Lamarck had been influenced by the idéologues who believed that matter had the inherent potential for some level of sensation and, therefore, life and ideas could emerge simply from matter, through something like spontaneous generation (Desmond 1992). For Lamarck, who saw life as more of an organizational property of matter, once very basic life emerged the subtle nervous fluid it contained could carve out new pathways, resulting in new forms of organization and, therefore, new forms of life (Desmond 1992, 45). In an ideal world, such actions would result in a linear development of life from simple organisms to humans. But, given that life was also subject to external stimuli, the environment could also play a role in shaping the development of life. The idéologues had argued that animals would have to adapt to a changing environment. For Lamarck, this meant that in a changing environment certain organs would be used in different ways and this would lead to changes in the pathways being carved out by the nervous fluid and a changing structure for the animal. Most importantly, these changes would be inherited by the next generation, leading to Lamarck's famous concept of the "inheritance of acquired characteristics."

One of Lamarck's leading supporters was Etienne Geoffroy Saint-Hilaire whose goal was to discover a "unity of plan" that governed the animal kingdom (Desmond 1992, 47). More than Lamarck's, his work had great appeal in medicine in both France and Great Britain. Unlike Cuvier who tried to explain animal structure totally in terms of the function of various organs, Geoffroy searched for similarities, or homologies, between different animals so as to be able to discern some master plan governing all animals (Desmond 1992, 51). This unity of plan did not govern function but represented a transcendent plan, and Geoffroy saw it as the job of the anatomist to search out and discover this plan. Philip F. Rehbock has argued during the early nineteenth century much of natural history in Great Britain, and especially Scotland was becoming "philosophical" or "transcendental" (Rehbock 1983, 3–12). Rather than simply classifying species in order to show the great variety of God's creation, the philosophical naturalists sought to discover transcendental laws or plans that governed all living things. Much of this approach was influenced by German idealist philosophy represented by the work of Immanuel Kant, J.W. Goethe, and *Naturphilosophie*, or by the British philosophy of the Cambridge Neo-Platonist or the writings of Samuel Taylor Coleridge (Rehbock 1983, 9). Geoffroy's unity of plan set the stage for a transformist view of the development of life. Rather than focusing on the innate conditions that led to the linear development of life from simple organisms to humans, as Lamarck had, Geoffroy placed more emphasis on the role of the environment, which had always been simply a secondary mechanism for Lamarck.

According to Jenkins, many of these Lamarckian and Geoffroyian ideas are present in the "Observation" article that Secord attributes to Jameson. The author of the article notes that the fossils of more tropical animals are often found in colder regions of the globe, which implies that the earth's climate has undergone change (Jenkins 2016a 539). The author then raises the question if climate change might modify the development of life. Using evidence from "artificial selection," or "scientific breeding" in agriculture, the author makes the following argument concerning species:

> But are these forms as immutable as some distinguished naturalists maintain; or do not our domestic animals and our cultivated or artificial plants prove the contrary? If these, by change of situation, of climate, of nourishment, and by every other circumstance that operates on them, can change their relations, it is probable that many fossil species to which no original can be found, may not be extinct, but have gradually passed into others.
>
> (Jenkins 2016, 539)

Wilson makes several comments that would lead one to believe that Jameson was teaching many of the ideas that were put forward by Lamarck, and especially Geoffroy. After noting that many of the displays in Jameson's Natural History Museum were simply stuffed animals, Wilson goes on to say:

> No one of the great naturalists in reality belonged to this skin-deep school. They knew that peculiar external characters were the result or counterpart of peculiar internal structure. They knew that systems of classification were to a great extent only matters of convenience, and that they had scientific value only in so far as they recognised all the affinities, analogies, and relationships which linked together the objects of classification.
>
> (G. Wilson and Geikie 1861, 114)

This seems very similar to Geoffroy's idea of unity of plan. An example of what might be considered the emergence of a philosophical or transcendental approach to mineralogy, Wilson says of Jameson:

> He had seen the concrete crystallography of his early days, which built up a great cube out of a multitude of infinitesimally small one, and acknowledged an infinity of primary forms, replaced by an abstract scheme, dealing in ideal axes, arranged in a very few sets or systems, each of which supplied a sufficient scaffolding for a number of geometrically related forms.
>
> (G. Wilson and Geikie 1861, 116)

As we will see in the next chapters, there is a great deal of evidence that Wilson was influenced by many of Jameson's ideas. In a great number of his publications and lectures, Wilson refers to "unity in variety," "unity of organization," "unity of nature," "mighty plan," "one ideal archetypal form," or simply the word "unity." Just as many times Wilson makes an argument that climate change can influence the development and the distribution of life on the earth. Finally, in his lecture "On the Physical Sciences which form the Basis of Technology," in two different places he refers to the ability of domesticated animals to be shaped by a form of selection (G. Wilson 1857, 97–99). Finally, Jameson may have influenced or strengthened Wilson's monogenesis beliefs. Both George and his brother Daniel believed that all human beings had a common origin rather than believing that each race was separately created (Swinney 2016, 182). In an article, Bill Jenkins provides an undated drawing from a student's notes from Jameson's natural history lectures (Jenkins 2016b 436–440). The drawing shows the relationship between the races, with Caucasians being at the top and then two lines descending from the Caucasians, one line leads to Mongols and then to Americans (assumed to mean Native Americans), and the other line leads to Malays and Negroes. Jenkins notes that this is essentially similar to the ideas of Johann F. Blumenbach.

In looking back on his days as a student, Wilson saw two "transforming influences" that took place in biology and chemistry (G. Wilson and Geikie 1861, 118–124). He argued that the microscope had transformed biology. Although microscope dates back to at least 1590 and had been used by Robert Hooke in his *Micrographia*, it was not widely available for use in biology. The microscope opened up new ways to study the animal and vegetable world and created new fields of science, such as histology. Possibly, anticipating his idea that the industrial sciences should be symbolized by an open hand with an eye in the palm, Wilson saw the microscope bringing about an interaction between the eye and the hand. He said, "an improved tool improves the skill of the hand, and here also the eye which, together with the hand, uses it" (G. Wilson and Geikie 1861, 121). The other transforming device was the organic combustion-tube that opened up new ways to analyze organic materials. Wilson argues that Robert Hooke had the idea that substances could be analyzed by burning them, but the beginning of organic analysis depended upon the discovery that water was a combination of hydrogen and oxygen, and that carbonic acid (what we now call carbon dioxide) was a combination of carbon and oxygen. Given this knowledge organic materials that consisted of only carbon, hydrogen and oxygen could be analyzed by burning them. The carbon would be converted into carbonic acid, and hydrogen into water, and then from the weight of the water and carbonic acid the amounts of hydrogen and carbon could be calculated and the weight of oxygen could be determined from the difference between the weight of the sum of carbon and hydrogen and the weight of the original compound. With the combustion-tube, the analysis of organic compounds went from days of work to days and even minutes of work.

While the microscope and the combustion-tube transformed biology and chemistry through the use of the eye and the hand, the stethoscope began to transform medicine through the use of the ear (G. Wilson and Geikie 1861, 131–133). The stethoscope had been invented in 1816 by René Laennec in France. Before this, doctors had to shake bottles partially filled with different liquids in order to diagnose the sounds coming from the chest. Wilson would see the stethoscope important enough that in 1847, he wrote a poem in honor of it (J. Wilson 1866, 220–22).

Influence of Philosophy

Wilson's education and training did not take place only in the lecture hall. His sister notes that George began participating in the Edinburgh Zetalithic Society, which he organized with his cousin James Russell, a student of William Hamilton, to develop reasoning faculties and the facility of speaking in public (J. Wilson 1866, 17). She also notes that in December of 1835, Wilson began keeping a journal of his thoughts and readings which his sister thinks show the "metaphysical bent of his mind" (J. Wilson 1866, 21–29). Most of the entries have to do with what was called the science of mind but that we would call psychology. For example, he records an unusual dream and that upon waking having a "sensation of being alone in a great hall or boundless valley" gave him a feeling of both loneliness and happiness along with an idea that "some invisible being of great power" was the cause of the feelings (J. Wilson 1866, 22). He goes on to attribute the dream to the fact the day before he had been reading Thomas De Quincey's "Confessions of an Opium-Eater" as well as having a lively conversation with an intellectual and imaginative friend which led him to a feeling of great excitement and attributes his dream to the role of his imagination being stimulated. In another example of the role of the relationship between the emotions and the mind, he relates a story of hiking by an old castle near the Tweed and deciding to jump through a window to examine the interior. As a result, he found himself knee-deep in mud. His immediate reaction was one of laughter but later when thinking about the incident, he had feelings of fear and horror thinking about the fact the mud could have been much deeper and could have swallowed him up. He concludes from this experience:

> how false the common idea is, that what causes of joy or grief are over, the effects will cease; but in all minds of any power, both will be immeasurably increased by reflection deepening their hues and heightening their effects, and producing deep and ineffaceable impressions on the heart of the thinker.
>
> (J. Wilson 1866, 27)

In another entry, Wilson reflects on his feelings as a boy to that of a young adult. As a boy, his pleasures resulted from an animal consciousness of simply being alive, but with the development of both his body and his brain, his pleasures came from objects that were beautiful and as a result elicited strong emotions. But along with pleasure in beauty, Wilson also noted that he also began to think of sad subjects and

> the prospect of evil and misery, and sin and woe, affects me much more powerfully than it did of old. In short, now my mind is much more developed than two years ago, and can ascend and descend much more widely than it could at that time, and my joy and sorrow is much more the result of legitimate causes than it was then.
>
> (J. Wilson 1866, 28–29)

In his diary entries, Wilson refers to two authors concerned with the philosophy of mind, and looking at these authors may provide another influence on the development of Wilson's thought (Tressler 2013; Glance 2001). In his entry on dreams, he mentions the work of Robert Macnish, who was trained in medicine but turned to literature and philosophy of mind (J. Wilson 1866, 23). His most famous work was *The Philosophy of Sleep* published in 1830 with a revised edition in 1834 (Macnish 1830). In his entries on the emotions, Wilson mentions of work of John Abercrombie, who was also trained in medicine, becoming the king's physician in Scotland and medical advisor to Sir Walter Scott, but he turned to the philosophy of mind (J. Wilson 1866, 25). He was the author of *Inquiries Concerning the Intellectual Powers and the Investigation of Truth* (1830) and *The Philosophy of the Moral Feelings* (1835) (Abercrombie 1830; Abercrombie 1833). It is not clear which edition of Macnish's book that Wilson read but it is more likely the 1830 edition since that edition makes more references to Scottish Common Sense philosophy, and the 1834 edition is based more on phrenology. Even though phrenology was being widely discussed in Edinburgh during the 1830s and 1840s, Wilson never mentions its theories in any of his works. In fact, there is reason to believe that Wilson opposed phrenology since much of the criticism of the anonymously published work *Vestiges of the Natural History of Creation* was aimed at its connection to phrenology and as we will see in the next chapter, Wilson led some of the opposition to the work (Secord 2000, 284–286). In any case, the chapters on dreams in each edition are quite similar with the exception of more references to Common Sense in the 1830 edition and one brief reference to phrenology in the 1834 edition. Macnish explains dreams as an aberrant state since the brain should be at rest but if one or more of the faculties is awake, dreams will take place (Macnish 1830, 50). A very interesting passage in Macnish's book discusses how dreams can change the perception of space and time and gives the following example:

> This curious psychological fact [changing perceptions of space and time], though occurring under somewhat different circumstances, has not

escaped the notice of that singular and highly-gifted writer, 'The English Opium-Eater.' – 'The sense of space,' says he, 'and in the end the *sense of time* were both powerfully affected. Buildings, landscapes, &c., were exhibited in proportions so vast as the bodily eye is not fitted to receive. Space swelled, and was amplified to an extent of unutterable infinity. This, however, did not disturb me so much as the expansion of time. I sometimes seemed to have lived for seventy or a hundred years in one night; nay, sometimes had feelings representative of a millennium passed in that time, or, however, of a duration far beyond the limits of human experience'.

(Macnish 1830, 62–63)

This statement from Macnish's book is very interesting since as we have seen above Wilson makes a reference in his diary to having a very vivid dream about a boundless valley after reading "Confessions of an Opium-Eater." This seems to closely resemble De Quincey's description of how space could be swelled under opium. It is not clear what connection there is between Wilson's reference to "Confessions of an Opium-Eater" and Macnish's reference. It might be that Wilson read Macnish's book and then decided to read De Quincey and subsequently had the dream he recorded, but it is clear that he read Macnish (both editions have almost the exact same story about the "Opium-Eater"). If Wilson were reading Macnish's book, he may have been influenced by other ideas. A key element of Macnish's theory of dreams is that they can be influenced by things that affect the body, such as food, different bedclothes, an arm or leg falling out of the bed, smoke, and sounds (Macnish 1830, 57–62). This might have encouraged Wilson to think about the relationship between the senses and the mind which was the focus of much of Scottish Common Sense philosophy. As we will see in a later chapter, Wilson was very interested in the senses and wrote a popular book on the five senses (G. Wilson 1853).

Wilson's reference in his diary to John Abercrombie is even more interesting since it could lead to more of a connection between Scottish Common Sense philosophy and Wilson's thinking. While not as widely known as Thomas Reid or Dugald Stewart, Abercrombie is seen as one of the leading Common Sense philosophers in the early nineteenth century (although following Stewart he abandoned the term "common sense" and replaced it with "first truths" or "primary principles" because of its ambiguity and possibility for misunderstanding) (Fieser 2000, vii). As we have seen, Common Sense philosophers shared a belief that humans had intuitive common sense beliefs about the world. Thomas Reid believed in a natural conjunction between things, or signs, and effects, or the signified. Common Sense philosophers were also focused on developing a philosophy of mind or a science of mind which might link together the senses and the mind. Abercrombie believed that his work in neurology could provide a scientific basis for a philosophy of mind. He divided mental phenomena into three categories: the simple intellect that perceives, remembers, and combines facts; the passive emotions that create pleasurable or painful feelings; and the active emotions that directly influence moral and social conduct (Abercrombie 1830, 13–14). Abercrombie asserts that the passive emotions may be influenced or excited either through our relations to other human beings, but they can be excited by material or inanimate objects. In the latter case, material objects can excite emotions of "the sublime, the beautiful, the terrible or the ludicrous" (Abercrombie 1830, 14). He goes on to connect the passive emotions excited by objects with the fine arts, including painting, sculpture, music, and poetry. This idea seems to be very close to Wilson's diary entry, in which he records how looking at objects could bring about strong emotions and in his contemplation of the beautiful, he specifically mentions the paintings of J.M.W. Turner and John Martin, the sculpture of Antonio Canova and beautiful poetry (J. Wilson 1866, 28). Wilson might have picked up another important idea from Abercrombie. In his *Inquiries*, Abercrombie discusses the relationship between science and art. He says:

> The object of all *sciences* is to ascertain these established relations of things, or the tendency of certain events to be uniformly followed by certain other events; in other words, the aptitude of certain bodies to produce, or to be followed by, certain changes in other bodies in particular circumstances. The object of *art* is to avail ourselves of the knowledge thus acquired, by bringing bodies into such circumstances as are calculated to lead to those actions upon each other of which we have ascertained them to be capable. Art, therefore, or the production of certain results by the action of bodies upon each other, must be founded upon science, or a knowledge of their fixed and uniform relations and tendencies.
>
> (Abercrombie 1830, 12)

As we will see in a later chapter, Wilson makes a very similar argument about technology or the industrial arts, which he argues must be based on science.

Also, in the *Inquiries*, Abercrombie provided a Common-Sense-based theory of the origins of our knowledge that may have also influenced Wilson's later thinking. According to Abercrombie, the origin of our knowledge first arises from our senses through which we gain knowledge of external things (Abercrombie 1830, 41). The sense impressions then become "the occasions on which the various powers of the mind are brought into action. These powers themselves then become the objects of consciousness or reflection …" (Abercrombie 1830, 40). Here, Abercrombie seems to be following not only the Common Sense idea of the importance of the sense but also the importance of certain innate aspects or powers of the mind. The use of these powers then leads through reflection upon the relations of the facts acquired through our senses we acquire notions of "time, motion, number, cause and effect, and personal identity; and we acquire farther the impression of certain fundamental laws of belief, which are not referable to any process of reasoning ..." (Abercrombie 1830, 41). Abercrombie's idea that knowledge originates in perception and reflection might have had an influence on Wilson, since the main purpose of Wilson's diary:

> is the wish to treasure up the prominent features of my mind as it acts at present, both to watch its progress, and to afford a fund of pleasing delights afterwards, in musing over the thoughts of my younger days ....
>
> (J. Wilson 1866, 21)

But Abercrombie may have also stimulated Wilson's thinking in another way. Abercrombie says:

> But in point of fact, the knowledge which is acquired by an individual, through his own perception and reflection, is but a small part of what he possesses; much of the knowledge possessed by every one is acquired through the perception of other men.
>
> (Abercrombie 1830, 41)

Abercrombie refers to this as "testimony," so knowledge is composed of sensation and perception, consciousness and reflection, and testimony. As we will see, much of Wilson's career was aimed at creating knowledge through public lectures, what Abercrombie might have labeled as testimony.

The Making of a Chemist

Wilson completed his formal course work in 1837 and subsequently passed the examination of the Royal College of Surgeons of Edinburgh, but he was not old enough yet to obtain his M.D. degree. In order to obtain more knowledge in chemistry, he became an assistant in the laboratory of Robert Christison, with whom he had studied Materia Medica (J. Wilson 1866, 41–42). Another assistant in the laboratory was Samuel Brown who would become one of Wilson's closest friends. Christison's research on toxicology provided Wilson with new excitement about chemistry and new ways of thinking about it. He seems particularly fascinated by the idea that chemists could artificially produce many of the same medicinal compounds that were produced naturally in plants (G. Wilson and Geikie 1861, 128). This idea that chemistry could be based on biology would play an important role in Wilson's future thinking, especially in the paper/lecture "On the Physical Sciences which form the Basis of Technology," which we will discuss in a later chapter. But Wilson went a step further and said:

> Besides extracting the quintessences which Nature had so diligently hidden and diluted through the mass of each medicinal plant, as well as infringing, or, as it were, breaking Nature's patent for manufacturing them, might it not be possible to produce compounds which, though no

plant yielded them, and for manufacturing them none guarded a secret or claimed a patent, were nevertheless as powerful in their action as any natural acid, alkaloid, or oil?

(G. Wilson and Geikie 1861, 129–130)

That is, not only could chemistry be based on biology, but a new artificial biology might be based on chemistry.

Wilson's growing interest in chemistry led him to make an important career decision. On October 6, 1837, he writes a rather coy letter to Margaret Mackay who would become his brother Daniel's wife. After discussing some of his experiments in Christison's laboratory and some of his recent absent-mindedness, he goes on to say:

Now I think I know the reason of all this mental absence ... I am over head and ears in love, and the object of my attachment so thoroughly engrosses my thoughts, that I have scarce a speculation to give anything else ... If you wish to see the birth, descent, and fortunes of the family, I would refer you not to Burke's Peerage, but to the Encylopaedia where under the article 'Sciences,' you will find a minute history of the family; and if you ask me which of the daughters has awakened in me such admiration, I reply, the 'Right noble Science of Chemistry,' who in my eyes is by far the most attractive and interesting of the family.

(J. Wilson 1866, 44–45)

In an interesting note at the end of the letter, Wilson tells Miss Mackay that if she wants to know more about the subject, she should read Mary Somerville's *On the Connexion of the Physical Sciences*. As we have noted in the previous chapter, Somerville's book reflected a new unified vision of natural philosophy.

The fact that Wilson had been reading Somerville's *Connexion* can give us some important insights into how his thinking is developing. As we have already seen, Jameson's course may have led him to think about connections between geology and biology, and his work in Christison's laboratory was leading him to think about the connections between chemistry and biology, but the fact that he has been reading Mary Somerville's book indicates that he is taking a much broader view of the connections between the sciences. Mary Somerville was born in Scotland and became attracted to mathematics and science at age fifteen. Much of her learning was through tutors but she also had the help of mentors such as Lyon Playfair and William Wallace in Edinburgh. It should be noted that Wilson studied mathematics with Wallace while a student at the University. Somerville also had the advantage of learning science through friendships with many of the leading scientists of the time. While in London, she socialized with Charles Babbage, Michael Faraday, William Herschel, and Charles Lyell and translated Pierre-Simon Laplace's *The Mechanism of the Heavens*. This translation led her to meet Laplace's widow on a trip to Paris and she met some of the leading French scientists, such as André-Marie Ampère, François Arago, Jean-Baptiste Biot, and Joseph-Louis Gay-Lussac. While in Paris, she began writing *Connexion* which was published in 1834. The book, which she says in her dedication to the Queen, was "my endeavour to make the laws by which the material world is governed more familiar to my countrywomen" (Somerville 1840). The book was widely popular, and not just with women. It would go on to sell 15,000 copies, becoming one of the best-selling books on science before the publication of Darwin's *On the Origin of Species* in 1859. It received praise from François Arago, one of France's leading scientists, and from Alexander von Humboldt, one of Germany's leading scientists (Holmes 2014, 43). It also received a very positive review from William Whewell in the influential *Quarterly Review* in 1834. His review also became famous for being the first place where he suggested the use of the word "scientist" (after rejecting philosophers, savants, nature-poker, and nature peepers) to designate "students of the knowledge of the material world" (Whewell 1834, 59–60).

The book, almost 500pages long, covered almost everything associated with the physical sciences but excluded chemistry and biology. Although she does not set out a specific plan for the book, given its title it is clear that her main purpose is to explain the relationships that exist between the various branches of the physical sciences (Somerville, 1840, preface). Not surprisingly for someone who was born in Scotland, her approach draws on some of the ideas of Common Sense philosophy. Common Sense philosophy drew heavily on the ideas of Francis Bacon and Isaac Newton and attempted to bring together an experimentalist and mathematical approach to knowledge. In her introduction, she draws on both of these approaches. In a statement that reflects both Baconian inductivism and some of the Common Sense theories of knowledge that we have already discussed, she says: "Our knowledge of external objects is founded upon experience, which furnishes facts; the comparison of these facts establishes relations, from with the belief that like causes will produce like effects, leads to general laws" (Somerville 1840, 3). Somerville seems to believe that the most important general law, one that connects everything, is Newton's law of gravitation. She again says:

> Astronomy affords the most extensive example of the connection of the physical sciences. In it are combined the sciences of number and quantity, of rest and motion. In it we perceive the operation of a force which is mixed up with every thing that exists in the heavens or on the earth; which pervades every atom, rules the motions of animate and inanimate beings, and is as sensible in the descent of a rain drop as in the falls of Niagara; in the weight of the air, as in the periods of the moon. Gravitation not lonely binds satellites to their planet, and planets to the sun, but it connects sun with sun throughout the wide extent of creation, and is the cause of the disturbances, as well as of the order, of nature: since

every tremor it excites in any one planet is immediately transmitted to the farthest limits of the system, in oscillations, which correspond in the periods with the cause producing them, like sympathetic notes in music, or vibrations from the deep tone of an organ.

(Somerville 1840, 1–2)

As we will see in the next chapter, this vision of a grand unity in nature would appeal to Wilson, but there are other ideas in her book that he might have found appealing. For example, Somerville discusses how climate and temperature difference affects the distribution of plants, a topic similar to Jameson's discussion on how a cooling earth would have affected the distribution of plants and animals (Somerville 1840, 291). In the same chapter, Somerville argues for a common origin for all humans based on similarities in physiology and language. She says: "it appears from a comparison of the principal circumstances relating to the animal economy or physical character of the various tribes of mankind, that the different races are identical in species," and after discussing similarities of language around the globe that "it may be inferred, that the nations speaking the languages in question are descended from the same stock …" (Somerville 1840, 300). This might have appealed to Wilson's known monogenesis views. There is definitely one part of Somerville's book that we know that Wilson must have found interesting because he closely copies one of her examples in one of his later works. In discussing sound, Somerville says: "[t]he propagation of sound may be illustrated by a field of corn agitated by the wind" (Somerville 1840, 146). She goes on to explain that when the wind blows through a cornfield, the individual stalks of corn oscillate back and forth but the waves are not physically moving the stalks out of their fixed location but only oscillating them. Wilson in his "Sketch of the Life and Works of the Hon. Robert Boyle," published in 1849, says that sound can be understood by observing "the spectacle of a field of growing corn, shaken by a gentle wind," and goes on to note the "ears of corn, however, have not been swept from one corner of the field to the other," but each ear "has only moved forward a little space in the direction of the wind, and then moved back to its original position" (G. Wilson 1862b, 221). Finally, Somerville's book may have either stimulated or reinforced Wilson's interest in Alexander von Humboldt whom she mentions at least ten times and saw as creating a global connection of scientific research programs (Holmes 2014, 432–433). We will discuss Humboldt's influence on Wilson in the next chapter.

In 1837, Wilson's older brother Daniel moved to London to try to establish a career as an illustrator. In the winter of that year, George began teaching an informal course on the chemistry of nature at his father's house (Balfour 1860, 4–5). The audiences for these lectures were women. He notes in his diary that "I meet with scarcely one lady in ten or fifty, who has sufficiently cultivated her natural intellectual powers. This winter shall see me do my utmost to suggest an improvement among my own small circle" (Balfour

1860, 4–5). Wilson notes that the lectures were received with praise and encouragement, and he would continue to make a point of contributing to the education of women throughout the rest of his life.

In 1838, Wilson went to visit his brother in London. He seems to have been particularly interested in meeting Thomas Graham. Graham was born in Glasgow and studied medicine at the University of Edinburgh and then became Professor of Chemistry at Anderson's College before becoming Professor of Chemistry at University College, London. Graham's work on the diffusion of gases would lead to what is known as "Graham's Law," which states that the rate of diffusion of a gas is inversely proportional to the square root of its density. He would later do work that led him to be seen as the founder of colloid chemistry. Wilson had only intended to stay in London for a short visit with his brother, but after meeting with Graham, both in his laboratory and during dinner, Wilson was offered a position as an unpaid assistant in Graham's laboratory (J. Wilson 1866, 64–65). This opportunity gave Wilson more time to spend with his brother and also allowed him to work on the thesis that he needed in order to obtain an M.D. degree. The position also introduced Wilson to a group of rising stars in chemistry who also worked as assistants to Graham. The most important were Lyon Playfair, who would be influential in the later establishment of the Industrial Museum of Scotland, and David Livingstone, who would become the famous African explorer and would provide objects from Africa for the Museum (J. Wilson 1866, 66). While working in London, Wilson attended some of Graham's lectures, assisted with some classes on practical chemistry, and assisted Graham in his research on the "Laws of Combination," between different substances which were probably connected to his research on the diffusion of gases, but he spent most of his time working on his thesis for his M.D.

There did not seem to be any prospects of permanent employment in London, so after spending about five or six months in Graham's laboratory, Wilson returned to Edinburgh in April 1839 to complete the requirements for his M.D. which he did in June of 1839. His thesis was on haloid salt in solution. It involved a complicated question of whether salt dissolved in water remains a chloride of sodium or becomes a hydrochlorate of sodium (Gladstone 1860, 510). With the state of knowledge at the time, Wilson was convinced that he had solved the problem but later in 1855 he had to raise questions about the conclusiveness of his results. In any case, his thesis was nominated for a prize and in the autumn, he presented his results at a meeting of the British Association for the Advancement of Science (BAAS) in Birmingham. At the meeting, Wilson participated in the first of the "Red Lions" dinners which became a tradition at all future BAAS meetings. Given the costs of attending the official banquets, especially for younger members, Edward Forbes invited some other naturalists to dine with him nightly at the "Red Lion," a small tavern where he was staying (J. Wilson 1866, 91). The dinners were so pleasant that the diners decided that there should be a Red

Lion dinner at all future BAAS meetings and Wilson almost never missed one of the dinners.

Influence of the Brotherhood

Also, in 1839, Wilson was inducted into The Universal Brotherhood of Friends of Truth that had been created by Edward Forbes while he was a student at Edinburgh (J. Wilson 1860, 223–231). This Masonic-like organization had its origins in January of 1835 from a group of students who edited the *University Maga*, a weekly satirical magazine, and labeled themselves the *Maga* Club, a group of medical, literary, and divinity students led by Charles Erskine Stewart (G. Wilson and Geikie 1861, 190–202). The original purpose of the Club was simply good fellowship and at first, it mainly focused on drinking with the club nights beginning with nine toasts and ending with singing "Rule Britannia." The magazine's main focus seemed to be that of satirizing professors through drawings and poems. Only twelve issues of the magazine were published, but it served to create a more noble fellowship. With the end of the magazine, the members of the Club, or the "Magi," established a brotherhood aimed at mutual assistance and encouragement of future careers. The motto of the new Brotherhood was Oinos, Eros, Mathsis (wine, love, learning). The symbol of the Brotherhood was a silk ribbon with the black letters O.E.M. worked into the fabric. The ribbon was worn across the chest which was to be worn at all times. During meetings, a small silver triangle with the Greek words on each side was worn at the end of the ribbon. While it was sometimes referred to it as the Oineromathic Brotherhood by 1838, it was named The Universal Brotherhood of Friends of Truth (J. Wilson 1860, 226). Forbes came to be the leading force of the Brotherhood, and in 1838 Forbes published a set of aims. Particularly important, especially in shaping some of the future ideas of some of the members, was the statement:

> Every step onwards we take in science and learning tells us how nearly all sciences are connected. There is a deep philosophy in the connexion yet undeveloped – a philosophy of the utmost moment to man; let us seek it out.
>
> (J. Wilson 1860, 226)

The Brotherhood was not seen as simply a student organization but a lifetime commitment. Members established chapters in France, Germany, England, and India (J. Wilson 1860, 223).

Wilson was proposed for membership in the Order by his friend, the chemist Samuel Brown, and joined an elite group of scholars. Of those in the Brotherhood, six went on to hold chairs at the University, including John Hughes Bennett, John Blackie, John Goodsir, Lyon Playfair, Edward Forbes, and George Wilson. There were also members who distinguished themselves outside of Edinburgh, including Wilson's friend Samuel Brown and Wilson's brother Daniel. Samuel Brown, Edward Forbes, and John Goodsir would all remain close friends with Wilson and probably had the greatest impact on his thinking. Forbes was probably the greatest influence on Wilson. They remained friends until Forbes's death and kept a regular correspondence during the time Forbes was in London, and as noted earlier, Wilson would write the *Memoir of Edward Forbes* (although he died before completing it). Forbes would become one of the leaders of the so-called philosophical or transcendental naturalists (Rehbock 1983, 68–75). Forbes studied comparative anatomy in an extra-mural course given by Robert Knox in Edinburgh. Before getting caught up in the Burke and Hare scandal, Knox was seen as the leading exponent of philosophical naturalism (Rehbock 1983, chap. 2). Influenced by the ideas of Immanuel Kant, the German idealist *Naturphilosophie*, and the transcendental anatomy of Geoffroy St. Hilaire, Knox believed that the goal of comparative anatomy was to discover the transcendental archetype that governed species. Following Geoffroy and Goethe, Knox saw this archetype as representing a "unity-of-plan," or a "unity-of-organization." He said: "A great plan or scheme of Nature exists, agreeably to which all organic forms are moulded" (Rehbock 1983, 49). Knox also believed that the archetypes were organized into what could be called the "great chain of being," although he did not believe that the chain of being was temporalized in any way and, therefore, rejected any idea of evolution (Rehbock 1983, 49–50). Finally, drawing from *Naturphilosophie*, especially the work of Lorenz Oken, Knox argued that the final form of an individual organism was the result of the conflicting, or polar, forces represented by unity-of-plan and diversity. He said: "The laws regulating the growth of *specific* forms are the antithesis of the laws presiding over the transcendental forms; the one bestows individuality on the species, the other struggles to reduce all to one type..." (Rehbock 1983, 52).

Edward Forbes also studied natural history with Robert Jameson whom we have already discussed and what he called his "philosophy of natural history" reflected both of their theories (Rehbock 1983, 69). He would later go on to say: "the transcendental philosophy of natural history [is] one of the most important developments of that science" (Rehbock 1983, 72). Forbes's philosophical or transcendental naturalism was probably influenced as much by Plato as it was by Kant (Rehbock 1983, 73; Desmond 1992, 79). Forbes probably received his Kantian ideas through the writings of Coleridge and his Platonic ideas through the writings of Henry More, one of the leading Cambridge Neoplatonists. Forbes's Platonism can be clearly seen in his interest in the trinity of things. Having been born on the Isle of Man, whose flag contains three legs in a triangular arrangement, he would often sketch the symbol (G. Wilson and Geikie 1861, 198). The identifying badge of his Brotherhood was a triangle that was founded on the ninth day of the third month; it had nine ceremonial officers and its meetings were at three minutes past nine. This interest in trinities could also be seen in a set of lectures Forbes gave with Samuel Brown whom we will discuss later. One of those lectures was titled: "Threefold constitution of the animal organism – Substance, Form and Intellect" (Rehbock 1983, 69).

In his later work, Forbes discussed two other topics that were important in transcendental natural history – unity-of-plan and polarity, both of which he seems to trace to the works of Goethe (Rehbock 1983, 70–73). Forbes's unity-of-plan can be seen in his treatment of genus, species, and individual (Rehbock 1983, 72–73). Species are made up of a group of individual organisms that share some common and unchanging characteristics and a genus was a related group of species. But for Forbes, these distinctions were not simple classifying devices. To quote his biography:

> But Forbes loved to deal with the organic world as a world of type and symbol – an embodiments of the thoughts of the Creator. A genus was to him a Divine idea that existed in its perfect form in the mind of God alone, and was only dimly shadowed forth to us. A species he regarded as the visible and individual, though partial manifestation of a generic idea ….
>
> (G. Wilson and Geikie 1861, 547)

But this divine plan also involved polarities. According to P.F. Rehbock, in a series of papers on *Sertularia* and *Plumularia*, Forbes interpreted that the development of the reproductive structures corresponded to a "preconceived, geometrically idealized pattern," and goes on to note that the "pattern involved two opposing axes: a 'vertical' growth axis and a 'spiral' reproductive axis" (Rehbock 1983, 70–71). The concept of polarity arose in what Forbes called his study of "Geo-Zoology" or Zoo-Geology" which involved studying the relationship between animal forms, fossils, and geological strata (G. Wilson and Geikie 1861, 320). In attempting to classify a newly established class of fossil, he noticed that it seemed to fit between more complex forms and simpler forms (Rehbock 1983, 103–113). Since Forbes was not a transformist but believed in a series of special creations, he did not interpret the relationships in evolutionary terms. Rather, he saw it as representing a struggle between simplicity and complexity, but this did not lead to linear progression. He noticed that in the fossil record, the greatest diversity existed at the beginning and end of each era which he saw as an example of polarity. According to Rehbock, Forbes seemed to attribute the decreased diversity during the middle of an era to the fact that each special creation was dependent upon the fact that "[s]uitable conditions have been met by the creation of suitable types" (Rehbock 1983, 109). That is, polarity drove both the origins and the diversity of life through interactions between a series of special creations and a changing environment and both of which were seen as manifestations of a divine idea (Rehbock 1983, 111). As we will see in the next chapter, Wilson seemed to be influenced by a number of Forbes's ideas.

Another member of the Brotherhood who would influence Wilson was John Goodsir. He began studying dentistry but then attended the University of Edinburgh Medical School and would go on to become Conservator of the Museum of the Royal College of Surgeons in Edinburgh, Curator of the University Natural History Museum, and finally Professor of Anatomy at the University of Edinburgh. Goodsir and his brother Harry (also a member of the Brotherhood) were close friends with Forbes and all three roomed together during 1839–1841 in Edinburgh. He would also conduct research and teach classes with Wilson (J. Wilson 1866, 124–126, 164). Like Forbes, Goodsir had also taken Knox's comparative anatomy course and had developed an interest in Coleridge while earlier studying humanities at St. Andrews University and later he became influenced by the anatomy of Goethe and Geoffroy St. Hilaire (Rehbock 1983, 91–98). Reflecting the goals of transcendental naturalism, Goodsir sought to discover the mathematical/geometrical forms that governed animal anatomy in a similar fashion to Kepler's formulation of the laws of planetary motion (Rehbock 1983, 94). But just as Newton had discovered the law of gravitation that explained Kepler's geometrical laws, Goodsir suggested that a similar law might govern the generation of animal forms. He notes:

> Newton had shown in his *Principia* that if attraction had generally varied as the inverse cube instead of as the inverse square of the distance, the heavenly bodies would revolve, not in ellipses but in logarithmic spirals, rapidly diffuse themselves, and rush off into space. It would be curious that if the law of the square were the law of attraction, the law of the cube might therefore prove to be the law of production.
>
> (Rehbock 1983, 94)

Rehbock suggests that Goodsir's idea for the production of animal forms is governed by a law of the third power, which may have arisen from his association with Forbes's Brotherhood and its idea of a mystical triangle (Rehbock 1983, 95–97). Rehbock goes on to note that Goodsir went on to develop a complete morphology based on the triangle or tetrahedron in which the human body could be seen to have a triangular outer shape and its inner shape could be divided into a series of triangles intersecting at various parts of the body (Rehbock 1983, 96). As we will see, Goodsir's transcendental anatomy may have influenced some of Wilson's ideas about the importance of unity of plan, or unity of organization. Rehbock also argues that Goodsir was influenced by the philosophy of William Hamilton, who at the time was trying to reconcile Kantian ideas with those of Common Sense philosophy (Rehbock 1983, 95). He quotes Goodsir as saying that: "Every sound intellect is necessarily – that is, is instinctively regulated, more or less, by the Laws of Thought" (Rehbock 1983, 95). Wilson was also most likely aware of the work of Hamilton since his cousin James Russell, who lived with the Wilsons and was close friends with George, was one of Hamilton's star pupils

(J. Wilson 1866, 166). Also, the ideas stated by Goodsir seem very similar to the ideas that Wilson may have gotten from reading the works of John Abercrombie, as discussed above.

One final member of the Brotherhood who would influence Wilson was the chemist Samuel Morrison Brown. Wilson and Brown had been friends even before joining the Brotherhood. They had been in the same class at Edinburgh, and they shared a laboratory together and often traveled together to scientific meetings (J. Wilson 1866, 52, 75). Also, as we have seen Brown also served with Wilson as an assistant in Christison's laboratory. Like Wilson, Brown had no interest in practicing medicine, and also like Wilson his goal was to become a chemist. From a very early age, Brown supported what he referred to as "transcendental chemistry" (Rehbock 1983, 87). Just as the transcendental naturalists sought to discover an underlying plan in the natural world, Brown believed that there was some unitary plan behind the chemical world. He based this belief on his idea that God was the source of all matter and if God is one all matter must have an underlying unity. According to his obituary, he held that matter was "multiform and yet one; in a word, matter, when first willed, must have the unity of the Author: – 'One God, – one law, – one *element*'" (Rehbock 1983, 88). Goodsir's biographer summarized Brown's thinking as follows:

> [He] joined in the poetic exaltations of Shelley, the erudite and metaphysical views of Coleridge, and the transcendentalisms of Goethe. Versed in the abstract, the abstruse, and the alchemical past of the Bacons and Van Helmonts, and daily sifting the current doctrines of Lavoisier and Dalton, he longed for a higher analysis than had been obtained by Cavendish, Priestley, or Davy and the laying of a more permanent foundation for his glorious science....He was a profound thinker, with the hopes of a theoretical seer, heralding the time when the composite organic and inorganic worlds would be resolved by man to a simple element, and the subtle agencies of light, caloric, and magnetism to one entity.
>
> (Rehbock 1983, 90)

Based on his idea of unity, Brown believed that the fifty, or so, elements known at his time were composed of identical atoms and their differences were due solely to different configurations of those identical atoms. This led him to the further idea that one chemical element could be transmuted into another element if the underlying atoms were rearranged into a different configuration. After a series of experiments by 1841, Brown became convinced that he had converted carbon into silicon. In 1843, he tried to use his new "discovery" as the basis for an application to succeed Hope as Professor of Chemistry at Edinburgh which led to what has been called the "transmutation war," with high profile figures taking opposing sides. The famous German chemist Justus Liebig rejected Brown's experiments saying he was "totally unacquainted with the principles of chemical analysis" (Rehbock 1983, 224, n.122). Others, such as Hamilton, Whewell, and Carlyle, praised his work. Both Wilson and Forbes were supportive of Brown's work. Wilson spent time trying to prove Brown's results, writing in December of 1843: "The repetition of Dr. Brown's experiments has engrossed me day and night, and still occupies my time" (J. Wilson 1866, 163). Forbes was also very supportive of Brown's work on transmutation. In a letter to Wilson in November of 1843, Forbes tried to connect Brown's work to explanations in natural history. He says:

> How do you account for certain so-called simple bodies being found only in the oldest formations, and others in only the newest or newer? That which is present in a stratum must be either there by an act of creation, or be derived from some previously existing body; but if that which is present be not identical with any previously existing body, or a combinations of any two or more previously existing bodies, and yet not be admitted as a new creation, it must be a transmutation. Now, it seems to me that the idea of transmutation of certain original inorganic forms, or even of a unity into multiplicity, is a simpler idea than the that of repeated acts of creation of inorganic bodies [This last sentence Forbes wrote in red ink and noted that it "is perhaps too obscurely transcendental to weigh with everyone."].
>
> (G. Wilson and Geikie 1861, 341–342)

Both Wilson and Forbes wrote testimonials in support of Brown being appointed Professor of Chemistry at Edinburgh (Brown 1843, 16–19, 23–24). Along with talking about Brown's knowledge of chemistry and his teaching, Wilson says of the transmutation research: "As some Chemists have denied the reality of this discovery, and others have refused to consider even its possibility, I wish to state my conviction, that by different processes Dr. Brown has repeatedly transmuted carbon into silicon," and he goes on to say "it is difficult to over-estimate the importance of the truth thus revealed to us, …" (Brown 1843, 18). In Forbes's testimonial, he specifically points to connections between Brown's discovery and work in natural history. He says that if Brown's work on transmutation proves correct

> they at once explain certain very difficult points in Natural History, which have hitherto been inexplicable; and if his views of the transmutability of the elements in general be as well founded as they seem to be, they are fitted to solve some perplexed and important questions in zoology, botany, and geology.
>
> (Brown 1843, 23)

In the end, others were unable to repeat Brown's transmutation experiment and the Chair of Chemistry went to William Gregory but it seems clear that Brown's work and ideas, even if not experimentally verifiable, had an important impact on the ideas of both Wilson and Forbes. As we will see Wilson, like Brown, was interested in seeing a unity of plan in nature, searching for connections between the organic and inorganic, and was interested in transmutation for much of the rest of his life.

Scientific Research

By the middle of 1840, Wilson was ready to begin his career in chemistry. He had obtained his M.D. the year before but had already decided that he was "in love" with chemistry. He continued working in Christison's laboratory for a period of time. His brother Daniel found Wilson a position in chemistry at a small school in London, but Wilson decided to turn down the position even though it would have allowed him to be in London with Daniel. After looking into the position, he told his sister: "it was a shabby affair, both in respectability and pecuniary value, and all of my friends here advised by to have nothing to do with it" (J. Wilson 1866, 103). Luckily a short time later, a better opportunity presented itself in Edinburgh. Wilson received a license from the Royal College of Surgeons and Physicians to lecture on chemistry. He notes that he received great support from members of the Brotherhood (J. Wilson 1860, 245). This was an honor usually reserved for Fellows of the Royal College, but it was granted to Wilson who became their first lecturer in chemistry. His courses would count for medical school diplomas but not at first for University degrees. These extra-mural classes were at first organized under the title Extra-Academical Medical School, and for a brief period under the name Queen's College. Eventually, these extra-mural courses were able to compete for students at the University. The new position brought with it a building, close to the University, that had a lecture hall and a small laboratory.

By the end of 1840, George Wilson was becoming established as a chemist. He had begun teaching chemistry at the extra-mural school in Edinburgh and found himself quite busy. Writing to his brother, Daniel, he notes that he is teaching six days a week to a class of thirty-one students and in addition, he was teaching a class in practical chemistry and instructing private students (J. Wilson 1866, 116). He quickly became a very popular lecturer and Edward Forbes would later say: "Wilson is one of the best lecturers I ever heard, reminding me more of the French school than our humdrum English, and is a man of high literary taste, and great general knowledge" (J. Wilson 1866, 118). In 1843, Wilson expanded his teaching duties (J. Wilson 1866, 159–160). The Highland and Agricultural Society of Scotland appointed him a lecturer at the Edinburgh Veterinary College and he also began teaching at the School of Arts. Finally, he began a course of lectures to women at the Scottish Institution. In a letter to his sister, he summarized his teaching:

> I have twenty students at my ten A.M. medical class; forty at my twelve o'clock (three days a week) veterinary class; some hundred young ladies

at the Scottish Institution; and some two hundred stout fellows at the School of Arts.

(J. Wilson 1866, 160)

Wilson also began conducting a program of scientific research. (There is a list of George Wilson's publications in Jessie Wilson's *Memoir.*) Although with one exception, Wilson's fame and legacy was not the result of his scientific research, his work covered a wide range of subjects (Gladstone 1860, 509–522). Still interested in Samuel Brown's atomic theory, Wilson published two papers in 1844 on transmutation, but neither were able to confirm Brown's claim (G. Wilson 1844; G. Wilson and Brown 1844). He recognized that near-drowning victims might be saved if administered oxygen and published a paper on the subject in 1845 (G. Wilson 1845). Beginning in 1846 and continuing for eleven years, Wilson published a series of seven papers on fluoride. This work attempted to settle the question of why the bones of ancient fossils had high levels of calcium fluoride, while recent bones had only small amounts (Gladstone 1860, 511–512). One theory was that dissolved fluoride in the earth had simply been deposited in the bones. But another more interesting theory, especially to Wilson, was that fluoride was the result of the transmutation of phosphate of lime. Wilson was able to show that fluoride is not only present in plants that animals use for food, but that it also occurs naturally in many minerals, springs, and seawater. This led him to conclude that the large amount of fluoride in ancient fossils was the result of water filtering into the bones.

Wilson also conducted research on the chemistry of bleaching, the specific gravity of chloroform, meteors, and photography, but the one area of scientific research that gave Wilson lasting fame was his study of color blindness which we will discuss in a later chapter (G. Wilson 1855). As we will see later, it is said that this work played a role in his appointment as Director of the Industrial Museum of Scotland.

References

Abercrombie, John. 1830. *Inquiries Concerning the Intellectual Powers and the Investigation of Truth.* Edinburgh: Waugh and Innes.

Abercrombie, John. 1833. *The Philosophy of Moral Feelings.* London: J. Murray.

Balfour, John H. 1860. *Biographical Sketch of the Late George Wilson, M.D.* Edinburgh: Murray and Gibb.

Brown, Samuel. 1843. *Testimonials in Favour of Dr Samuel Brown, Now a Candidate for the Chair of Chemistry in the University of Edinburgh.* Edinburgh: Neill and Co.

Davie, George Elder. 1961. *The Democratic Intellect: Scotland and Her Universities in the Nineteenth Century.* Edinburgh: University of Edinburgh Press.

Desmond, Adrian. 1992. *The Politics of Evolution: Morphology, Medicine, and Reform in Radical London.* Chicago: University of Chicago Press.

Fieser, James. 2000. *Scottish Common Sense Philosophy: Sources and Origins*, vol. 5. Bristol: Thoemmes Press.

Gladstone, John Hall. 1860. "Estimate." In *Memoir of George Wilson*, edited by Jessie Aitken Wilson, 509–522. Edinburgh: Edmonston and Douglas.

Glance, Jonathan C. 2001. "Revelation, Nonsense or Dyspepsia: Victorian Dream Theories." Unpublished paper presented at the Northeast Victorian Studies Association Conference.

Holmes, Richard. 2014. "In Retrospect: On the Connexion of the Physical Sciences." *Nature* 514: 432–433.

Jenkins, Bill. 2016a. "Neptunism and Transformation: Robert Jameson and other Evolutionary Theorists in Early Nineteenth-Century Scotland." *Journal of the History of Biology* 49: 527–557.

Jenkins, Bill. 2016b. "The Platypus in Edinburgh: Robert Jameson, Robert Knox and the Place of the *Ornithorhynchus* in Nature, 1821–24." *Annals of Science* 73: 425–441.

MacGregor, George. 1884. *The History of Burke and Hare: And the Resurrectionist Times.* Glasgow: Thomas D. Morrison.

Macnish, Robert. 1830. *The Philosophy of Sleep.* Glasgow: W.R. M'Phun.

Olson, Richard. 1975. *Scottish Philosophy of British Physics, 1750–1850: A Study in the Foundations of the Victorian Scientific Style.* Princeton, NJ: Princeton University Press.

Rehbock, Phillip F. 1983. *The Philosophical Naturalists: Themes in Early Nineteenth-Century British Biology.* Madison: University of Wisconsin Press.

Secord, James. 1991. "Edinburgh Lamarckians: Robert Jameson and Robert E. Grant." *Journal of the History of Biology* 24: 1–18.

Secord, James. 2000. *Victorian Sensation: The Extraordinary Publication, Reception, and Secret Authorship of Vestiges of the Natural History of Creation.* Chicago: University of Chicago Press.

Somerville, Mary. 1840. *On the Connexion of the Physical Sciences*, 5th ed. London: John Murray.

Swinney, Geoffrey N. 2016. "George Wilson's Map of Technology: Giving Shape to the 'Industrial Arts' in Mid-Century Edinburgh." *Journal of Scottish Historical Studies* 36: 165–190.

Tressler, Elizabeth. 2013. "Ecstasy and Solitude: Reading and Self-Loss in Nineteenth-Century Literature and Psychology." PhD diss., Boston College.

Whewell, William. 1834. "Review of On the Connexion of the Physical Sciences by Mrs. Somerville." *Quarterly Review* 51: 54–68.

Wilson, George and John Crombie Brown. 1844. "Account of a Repetition of Several of Dr. Samuel Brown's Processes for the Conversion of Carbon into Silicon." *Transactions of the Royal Society of Edinburgh* 15: 547–560.

Wilson, George. 1844. "On Isometric Transmutation." *Edinburgh New Philosophical Journal* 3: 82–103.

Wilson, George. 1845. "On the Employment of Oxygen as a Means of Resuscitation in Asphyxia, and Otherwise as a Remedial Agent." *Transactions of the Royal Society of the Arts.*

Wilson, George. 1853. *The Five Gateways of Knowledge.* London: Macmillan.

Wilson, George. 1855 *Researches on Colour-Blindness.* Edinburgh: Sutherland and Knox.

Wilson, George. 1857. "On the Physical Sciences Which Form the Basis of Technology." The *Edinburgh New Philosophical Journal*, new series 5: 64–104.

Wilson, George and Archibald Geikie. 1861. *Memoir of Edward Forbes, F.R.S.* London: Macmillan.

Wilson, George. 1862a. "Life and Discoveries of Dalton." In *Religio Chemici: Essays*, edited by George Wilson, 304–364. London: Macmillan and Co.

Wilson, George. 1862b. "Robert Boyle." In *Religio Chemici: Essays*, edited by George Wilson, 164–252. London: Macmillan and Co.

Wilson, Jess Aitken. 1860. *Memoir of George Wilson.* Edinburgh: Edmonston and Douglas.

Wilson, Jessie Aitken. 1866. *Memoir of George Wilson*, new ed. London: Macmillan and Co.

3 Unity in Variety

Wilson's Theology of Nature

While Wilson received praise for his work on color blindness, ultimately his fame and legacy would not rest on his record of scientific research. John H. Gladstone, a friend and memorialist of Wilson, wrote: "While many of Dr. Wilson's contemporaries could pursue a train of research with greater ability, none perhaps could render the new truth thus obtained so attractive by copius imagery and varied illustration" (Gladstone 1860, 520). Gladstone referred to Wilson as an "expounder" which seems to be an appropriate label, since it was through his expounding or popularizing that he created his legacy and his fame. Of course, much of Wilson's legacy and fame were the result of his position as Regius Professor of Technology at the University of Edinburgh and as founding Director of the Industrial Museum of Scotland which will be discussed in the next two chapters. But even before taking up those positions, Wilson had established himself as a promoter of a unitary vision of science through a series of lectures and publications that reached wide audiences. His textbook on chemistry, which was part of the Chambers's Educational Course, was meant to introduce basic concepts of chemistry to a general audience and sold 24,000 copies in just nine years and his popular book, *The Five Gateways of Knowledge*, which will be discussed later in some detail, sold over 8,000 copies in its first three years (J. Wilson 1866, 190 and 303; G. Wilson 1856; G. Wilson 1850a).

Medical Crisis

Wilson's unitary vision of nature, which we will discuss in this chapter, was obviously influenced by his years as a student at the University of Edinburgh and through his close friends in the Brotherhood. But beyond the influence of transcendental naturalism and Common Sense philosophy, this vision became shaped by a defining period in his life when he began to face a series of medical crises, some of which were life-threatening. Just before beginning his first set of lectures in the extra-mural school in Edinburgh in the fall of 1840, he took a short excursion to Perthshire in order to do some hiking. While hiking he sprained his ankle, which did not seem to be a serious injury, but attending the British Association meeting in Glasgow a few days

DOI: 10.4324/9781003212218-3

later aggravated his injury leading to an abscess (J. Wilson 1866, 111–116). Again the injury did not seem serious but was again aggravated over the winter when he began to suffer from rheumatism. In the fall of 1841, Wilson visited his brother Daniel in London and began to suffer from an inflammation of the eye, which became worse after his return to Edinburgh. His problem of being able to walk and to see continued into March of 1842. He did have a brief period of relief but by May, the problem with his ankle was getting worse. By December, his foot was no better and he had to rely on opiates to deal with the pain. As his sister writes: "A crisis was again approaching in George Wilson's life more momentous than any hitherto considered" (J. Wilson 1866, 147).

In January of 1843, a decision was made that the only solution to Wilson's ankle problem was a partial amputation of his foot. The surgery was performed by James Syme, Professor of Surgery at the University and he was assisted by John Goodsir, Wilson's friend. At the time, there was no accepted medical use of anesthetics, such as chloroform or ether for surgeries, so Wilson would be forced to undergo the amputation without any anesthesia. He describes his experience as follows:

> Suffering so great as I underwent cannot be expressed in words, and thus fortunately cannot be recalled. The particular pangs are now forgotten; but the black whirlwind of emotion, the horror of great darkness, and the sense of desertion by God and man, bordering upon despair, which swept through my mind and overwhelmed my heart, I can never forget, however gladly I would do so.
>
> (J. Wilson 1866, 150).

Evangelical Conversion

Although Wilson would speak of a sense of "desertion by God," the entire experience seems to have set him in a new religious direction. Wilson had been brought up in a somewhat religious home. His parents had been Baptists and as a young man, Wilson flirted with Episcopalianism, but until his medical crisis, he did not seem to exhibit any strong religious feelings. But even the anticipation of the surgery and possible death seems to have awakened some new feelings in Wilson. His sister describes the change, saying:

> [T]hat one thing was yet wanting in George Wilson's life, he himself freely acknowledged. Not yet had a living faith in God been developed, not had he dedicated himself to Him who claims each man as rightfully His own. In this time of deep thought, however, he underwent a mighty change. He, himself always regarded it, as a time when a new life dawned in his soul ...
>
> (J. Wilson 1866, 148–149)

During this period, he established a lifelong friendship with John Cairns, then a divinity student at Edinburgh. Wilson had been introduced earlier to Cairns through his cousin John Russell who lived with the Wilson family and had been in a humanity class with Cairns, but the experience of the surgery brought them closer together and Wilson would afterward refer to Cairns as his "spiritual father" (J. Wilson 1866,150). His recovery from the surgery went well, but in the summer he experienced the first signs of what would later be discovered to be tuberculosis which would cause him continual problems for the rest of his life. This new affliction seems to have strengthened his commitment to religion since in the summer of 1844 he decided to be baptized (J. Wilson 1866, 174–175). Since his parents were Baptists, he had not been baptized as a child and it was up to Wilson to decide on baptism. About the same time, Wilson decided to join the local Congregational Church led by the Rev. Dr. W. Lindsay Alexander, who had been a student of Thomas Chalmers, the famous Scottish evangelical theologian. In a letter of grievance sent to Spencer Walpole, Secretary of State for the Home Department concerning the of the University Test Act requiring Scottish professors to be members of the Church of Scotland , Wilson explains his own religious affiliation as follows:

> To prevent any misunderstanding, let me further state, that I am a member of a Congregational Church. There are two sections of Congregationalists, Independents and Baptists, who differ as to the mode, the subjects, and the significance of baptism, but agree in other respects, in reference to doctrine and church government. I am a Baptist, but regarding the difference with respect to baptism as not a valid ground of separation between Christians who are at one in other matters, I am a member of a church, the majority of whom, including their minister, the Rev. Dr W.L. Alexander, are Independents.
>
> (G. Wilson 1852a, 8)

The Independent branch of the Congregational Church was much more evangelical, supporting missionary movements. As we will see, Wilson was closely involved with the Edinburgh Medical Missionary Society. After Wilson gained a renewed interest in religion, he often made mention of it in his popular scientific lectures and publications and in his letters to friends and family (Balfour 1860, 11). But with one or two exceptions, his lectures and publications were not overtly theological. In his memorial to Wilson, John Gladstone writes:

> The beauty of Dr. Wilson's discourses and writings depended not a little on his religion, and on his fine aesthetic taste. His quotations from Holy Scriptures, and references to spiritual things were frequent, not in the form of pious deduction dragged in uncomfortably at the end of a lecture, but as the natural reflections of a mind thoroughly embued with the love of God and man …
>
> (Gladstone 1860, 521)

Science and Religion

Given Wilson's strong religious beliefs, it might be tempting today to dismiss Wilson as having any important role in the development of science, but as a number of scholars have recently shown, the role of religion was an important element in the development of Victorian science. During the late Victorian period, there seemed to be a conflict between science and religion. Many people point to the publication of John Draper's *History of the Conflict between Religion and Science* (1874) as evidence of some fundamental opposition between science and religion. But if we look at the book's introduction, it becomes clear that the conflict Draper is writing about is not between religion in general and science but between Roman Catholicism and science. He in fact has some praise for other religions, saying:

> I have little to say respecting the two great Christian confessions, the Protestant and Greek Churches. As to the latter, it has never, since the restoration of science, arrayed itself in opposition to the advancement of knowledge. On the contrary, it has always met it with welcome. It has observed a reverential attitude to truth, from whatever quarter it might come. Recognizing the apparent discrepancies between its interpretations of revealed through and the discoveries of science, it has always expected that satisfactory explanations and reconciliations would ensue, and in this it has not been disappointed. It would have been well for modern civilization if the Roman Church had done the same.
>
> (Draper 1875, x)

Shortly after, Andrew Dickson White published his book *The Warfare of Science* which did provide some examples of Protestants inhibiting the development of science, but the vast majority of examples involve the Roman Catholic Church and various Popes attacking science (A. White 1896). Even though there was a great deal of criticism of White's sloppy scholarship, such as claiming the Catholic Church still believed in a flat earth at the time of Columbus, White's "warfare" model played a significant role in the way people thought about the relationship between science and religion until the second half of the twentieth century, but recent scholarship has led to questions concerning this model, and scholars of the Victorian period have raised questions about its appropriateness for understanding that period (Barbour 1990; Turner 1974).

In his Gifford Lectures, Ian Barbour, who possessed a doctorate in physics from the University of Chicago and a divinity degree from Yale University, argued that the conflict model of the relationship between science and religion was only one of four possible models (Barbour 1990, 3–30). Along with the conflict model, Barbour also argued that to avoid conflict, science and religion could be thought about as independent of one another, or that science and religion could be seen as engaged in a dialog with each other, or finally science and religion could become integrated together in some manner. Barbour himself favored either the dialog model or the integration model as being the most fruitful for understanding the relationship between science and religion and included under the integration model the ideas of natural theology and a theology of nature, which we will see played an important role in the early Victorian era.

Frank M. Turner has argued that the idea of a straight-line movement from religious beliefs to secular thinking during the Victorian era, common among mid-twentieth century scholars, prevented them from gaining a complete understanding of the complex role played by religion throughout the entire Victorian period (Turner 1993, 9–10). He notes that a rigid distinction between religious thinking and secular thinking makes it difficult to understand the resurgence of Roman Catholicism and Nonconformist ideas, the Oxford Movement, or the Tractarians, religious politicians, such as William Gladstone, the role of Methodism among the working classes, and the role of natural theology in science, and he goes on to argue that "new sensibilities" arising from social history and the history of science have led to a "serious questioning of the secular thesis" (Turner 1993, 4–21).

While Turner agrees that by the end of the nineteenth century, scientists such as Thomas Henry Huxley, John Tyndall, and Francis Galton seemed to be giving support to the idea of a fundamental conflict between science and religion, this was not a unanimous view. More importantly, during the early Victorian age, Turner finds a great deal of support for the idea that science could play an important role in religion by providing examples that could support the idea of a natural theology (Turner 1993, 177–178). As early as the Middle Ages, Christians were encouraged to study the "book of nature" in order to find signs, or "signatures" of the mind of God (White 1967, 1203–1207). But the more modern notion of natural theology can be traced to William Paley's publication of his book *Natural Theology* in 1802. Turner attributes the new support for a natural theology to a reaction against late eighteenth-century scientists, such as Joseph Priestly, James Hutton, and Erasmus Darwin, whose radical political thought seemed to be aligned with the French Revolution. In reaction, supporters of a natural theology sought "to demonstrate that the natural and social orders understood within a religious or eternal framework ordained political and social passivity and that scientific knowledge both natural and economic was politically and socially inoffensive" (Turner 1993, 105). John Hedley Brooke argues that because natural theology relied on nature as a way of understanding God, rather than on differing written scriptures, it could have an appeal across sectarian lines and provide some level of unity where there had been divisions (Brooke 1991, 198–199). Paley's book became famous for the argument that if you found and studied a finely made watch, you would be required to infer that it could not have come about by some random process but had to be the result of an intelligent designer. Therefore, by finding designs in nature, which was one of the goals of science, one could conclude the existence of some grand intelligent designer, or God. This was the so-called "argument from design;" that is, from the existence of designs in the world one could argue for the existence of God.

The Bridgewater Treatises

Some of the most influential works on natural theology were the *Bridgewater Treatises* (Turner 1993, 111–117). As part of his will, Francis Henry Edgerton, Earl of Bridgewater, who died in 1829, provided 8,000 pounds to the Royal Society of London to sponsor the publication of works "On the Power, Wisdom, and Goodness of God, as manifested in the Creation." Published between 1833 and 1836, the eight original *Bridgewater Treatises* were authored by such people as Thomas Chalmers, leader of the Church of Scotland, John Kidd, chemist and geologist, William Whewell, philosopher of science, Charles Bell, physiologist and neurologist, Peter Mark Roget, known for his later *Thesaurus*, William Buckland, geologist, William Kirby, entomologist, and William Prout, chemist. In response to Whewell's book, Charles Babbage, inventor of the difference engine, wrote an unofficial *Ninth Bridgewater Treatise*. Ironically, most of the authors actually argued for distinctions between religious ideas and scientific knowledge, with Buckland and Whewell claiming "that scripture is an inadequate guide to the study and understanding of nature" (Turner 1993, 111). Many, if not all, of the *Bridgewater Treatises* emphasized the providential nature of God in creating a world suitable for human beings and often in doing so gave support to the idea that modern Europeans were the natural result of God's providence (Turner 1993, 113–114). The works also gave support to the idea that competition and industrialization were also part of God's plan. Prout and Roget justified the cruelty resulting from the competitions that existed in the animal kingdom as justifying the competition taking place in the industrial world. Buckland in his discussion of the geology of coal seemed to imply that God had providentially designed the earth (and Great Britain), for future industrial development. Whewell argued that by providing an atmosphere surrounding the earth and human sense organs that could produce and receive speech transmitted through that atmosphere, God had provided the means for human communication and thus human society and thus moral development.

There was a significant amount of criticism of the *Bridgewater Treatises*. Robert Knox objected to the "ultra-teleology" of the works and mocked them as the "Bilgewater Treatises" (Rehbock 1983, 56). Turner makes the interesting argument that the *Bridgewater Treatises* actually had the opposite effect than they intended and in fact led toward secularization. He says:

> They laid the foundation for an image of humankind and commercial industrial society that did not necessarily require the presence of God or a religious setting for human life. That is to say, they presented an

image of humankind and human society for which alternative naturalistic explanations and rationalizations could be substituted for the religious ones upon which they drew.

(Turner 1993, 111–112)

Wilson seems to have had ambivalent feelings toward the *Bridgewater Treatises*. In a letter to a young friend on January 15, 1847, he recommends that he read the *Treatises*, especially those of Whewell, Roget, Buckland, and Bell (G. Wilson 1862d, 6). Peter Roget's work on *Animal and Vegetable Physiology Considered with Reference to Natural Theology* (1834) may have played a particularly influential role on Wilson since it includes the idea of unity and variety that Wilson would often use in his later writings. Peter Roget, who would later gain great fame as the author of the *Thesaurus*, had graduated in medicine from the University of Edinburgh and would go on to become one of the founders of the University of London (Desmond 1992 228–231; Rehbock 1983, 56–59). Philip F. Rehbock notes that Roget's *Treatise* begins and ends with typical arguments based on final causes, but in-between he puts forward a tentative transcendental theory of morphology (Desmond 1992, 229; Rehbock 1983, 57–58). Roget argued that along with the governing principle of adaption to function, two other laws played a role in animal and vegetable physiology. He says: "In every department of nature it cannot fail to strike us that variety is a characteristic and predominant feature of her productions ..." (Rehbock 1983, 57). This resulted in what he called the law of variety. But this variety needed to be kept under control, and Roget argued that the law of variety is "circumscribed within certain limits, and controlled by another law, ... that of *conformity to a definite type*" (Rehbock 1983, 58). He would also refer to this law of conformity to a definite type as "unity of plan" (Rehbock 1983, 58). That is Roget is arguing that along with the emphasis on final cause or adaption that was the typical focus of natural theology, there also needed to be the transcendental idea of unity of plan. As we will see, Wilson often used the ideas of unity in variety in his later writings, saying that: "Unity in variety is the voice of all nature ..." (G. Wilson 1862e, 274; G. Wilson 1859, 21). Also, as we will see later, Wilson in his lecture on the "On the Character of God" would use an argument similar to Roget's and argue that the final cause could be reconciled with unity of plan (G. Wilson 1856, 65–70).

On the other hand, both Wilson and Forbes were often critical of the *Bridgewater Treatises*. Forbes believed that God's existence was a given and did not need to be proved from a study of nature, and Wilson complained that natural theology often ignored the problem of evil, and that the "absence of anything like a resolute attempt to look this great problem of physical evil in the face" was a significant failure (G. Wilson 1862b, 30–31; G. Wilson and Geickie 1861, 547). Wilson made one exception to his criticism of the "beautiful volume," by Thomas Chalmers. It may be that the more Wilson became involved with evangelicalism the more limitations he saw in natural theology. Wilson's minister had been a student of Chalmers, and Chalmers audited Wilson's lecture on transmutation in 1844 (J. Wilson 1866, 174).

Evangelical Science: A Theology of Nature

Thomas Chalmers was the author of the very first *Bridgewater Treatise*, which he titled *On the Power Wisdom and Goodness of God as Manifested in the Adaptation of External Nature to the Moral and Intellectual Constitution of Man* (Chalmers 1835). Chalmers had studied mathematics at St. Andrews and then theology at the University of Edinburgh where in 1828, he became Professor of Theology. He would later become known as the leader of a group of evangelicals who would break away from the established Church of Scotland to form the Free Church of Scotland in 1843 in what was called the "Disruption," which was the result over concerns that the British government was intruding into the affairs of the Scottish churches (Smith 1998, 17–22). Although Chalmers agreed to author the first *Bridgewater Treatise*, it seems he had a number of reservations about the entire program of natural theology. In his early work on *The Evidence and Authority of the Christian Revelation* (1814), he seemed to have repudiated natural theology as speculative and unscientific (Topham 1999, 145). But, based on criticism of *The Evidence*, Chalmers reevaluated his attitude toward natural theology and in his *Bridgewater Treatise*, he attempted to develop a revised approach to natural theology (Topham 1999, 159–164). He believed that a study of nature would provide only limited knowledge about the characteristics of God. He said: "We hold that the material universe affords decisive attestation to the natural perfections of the Godhead, but that it leaves the question of his moral perfection involved in profoundest mystery" (Chalmers 1835, 1, 51). By this, he seems to have meant that a natural theology could inform someone about such things as God's unity, eternity, omnipresence, and omniscience, but it could not provide evidence of God's goodness, righteousness, and justice (Smith 1998, 18). He states that "[t]he laws of nature may keep up the working of the machinery – but they did not and could not set up the machine" (Chalmers 1835, 1, 40). As an example, he notes that a torture device might exhibit the characteristics of design but not goodness (Chalmers 1835, 1, 51–52). Here he notes: "that Paley, so full and effective and able in his demonstration of the natural, is yet so meagre in his demonstration of the moral attributes" (Chalmers 1835, 2, 102). For Chalmers, while a study of matter may provide evidence of the existence of God, it is through the study of the human mind that evidence of the moral character of God can be found (Chalmers 1835, 1, 52054). This led to the development of what Jonathan Topham has labeled a natural theology of conscience and what Chalmers simply labeled a theology of conscience (Topham 1999, 165–167; Chalmers 1835, 1, 90). In his study of the human mind, Chalmers draws on the ideas of Common Sense philosophy, especially the work of Thomas Brown, James Beattie, and John Robison, to argue that the way in which the mind perceives nature is not simply based on some sense experience but is shaped by innate or intuitive powers of the mind (Topham 1999, 151; Chalmers 1835, 2, 141–142). Given this intuitive aspect of the human mind, it could be argued that the human mind, especially its conscience, in some way reflects God's mind and, therefore, by studying the science of the mind, one could gain insights into the moral attributes of God such as justice and righteousness (Brooke 1999, 24).

Chalmers was critical of the fact that natural theology provided little proof of the moral attribute of God but accepted that it might play a role in making inferences about those characteristics. He said: "We hold that the theology of nature sheds powerful light on the being of a God; and that, even from its unaided demonstrations, we can reach a considerable degree of probability, both for His moral and natural attributes" (Chalmers 1835, 2, 285). But Chalmers also sees "the insufficiency of that academic theism, which is sometimes set forth in such an aspect of completeness and certainty, as might seem to leave a revelation of a gospel wholly uncalled for" (Chalmers 1835, 2, 286). He goes on to compare "the speculations of human ingenuity" with "the certainties of well accredited revelation" and says we must not "idolize the light or sufficiency of nature" (Chalmers 1835, 2, 287). He goes on to say that:

> two positions are perfectly reconcilable – first, of the insufficiency of natural religion; and secondly, the great actual importance of it. It is the wise and profound saying of D'Alembert, that 'man has too little sagacity to resolve an infinity of questions, which he has yet sagacity enough to make.' Now this marks the degree in which natural theology is sagacious – being able, from its own resources, to construct a number of cases, which at the same time it is not able to reduce. These must be handed up for solution to a higher calculus; … It is a science, not so much of dicta as of desiderata… It puts the question, though it cannot answer the question;… Natural theology, then, however little to be trusted as an informer, yet as an inquirer, or rather as a prompter to inquiry, is of inestimable service.
>
> (Chalmers 1835, 2, 287–289)

Here, Chalmers is arguing that natural theology can have some role as a handmaid, but it must ultimately depend on a "higher calculus," by which he seems to mean revelation. It seems here that Chalmers is moving from natural theology to what Topham calls a theology of nature.

For Chalmers, natural theology also had a role in dealing with the question of sin and evil, a topic not often addressed by other natural theologies which usually focus on the goodness and rationality of the world so as to show the goodness and rationality of the Deity. But Chalmers, possibly reflecting his Calvinistic roots and the evangelical belief in the natural depravity of humans because of original sin, often emphasized the "principles of destruction," present in the world (Bebbington 1999, 128–129; Smith 1998, 18). He notes that the "same light which irradiates the perfections of the divine nature, irradiates, with more fearful manifestation than ever, the moral disease and depravation into which humanity has fallen" (Chalmers 1835, 2, 284). But this provides another role for natural theology since "[h]ad natural theology been altogether extinct, and there had been no sense of law or lawgiver among men, we should have been unconscious of any difficulty to be redressed, of any dilemma from which we needed extrication" (Chalmers 1835, 2, 284).

Theistic Science

While a number of Victorian scientists were critical of natural theology, they were attracted to what scholars such as Matthew Stanley labeled "theistic science," which could also be labeled theistic naturalism or a theology of nature (Stanley 2015, 4). While related to natural theology, theistic naturalism had a somewhat different emphasis and goal, but the issue gets confused since during the mid-nineteenth century natural theology often was used to also refer to what might be called a theology of nature (Topham 1999, 144; Brooke 1974, 8–9). Topham argues that while natural theology was based on reason alone, a theology of nature was a body of theological beliefs "based on natural reason, on revelation, or a combination of the two" (Topham 1999, 144). Much of natural theology was based on the work of Paley and his argument *from* design. That is, the existence and character of God could be determined from the designs that could be found in nature. Theistic naturalism turned that idea around into an argument *for* design (Stanley 2015, 48). That is, instead of trying to prove God's existence, if one accepted that God existed, one could look to nature to see the designs that God had imposed on the world. In discussing the Scottish minister Robert Candlish, James Secord gives a specific example of this theology of nature when he says:

> If we start from nature, Candlish said in his famous sermon 'Paul Preaching at Athens,' we approach God as a dim and vague abstraction. If we begin with Christ, and then go into nature, 'we see the God of judgement blazing everywhere.'
>
> (Secord 2000, 274)

As we will see, this could be linked to transcendental naturalism, whose goal was to discover the archetypes and the unity of plan that existed behind nature, but theistic naturalism would go further and assume that the source of such a unity of plan was God (G. Wilson and Geikie 1861, 547). Theistic science, unlike natural theology, was not a handmaid to religion, rather it drew on religion as a way to do science. As such, it aimed to create a theology of science.

Uniformity

A fundamental element of both theistic and secular science was the belief in the uniformity of nature and the uniformity of natural laws (Stanley 2015, chap. 2). Stanley argues that John Herschel's *Preliminary Discourse on the*

Study of Natural Philosophy (1830) "shaped a generation of science" (Stanley 2015, 34). In this work, Herschel argued that "The only facts which can ever become useful as grounds of physical enquiry are those which happen uniformly and invariably under the same circumstances" (Herschel 1851, sec. 110, 119). For Herschel, this idea of the uniformity of nature, or the order of nature, arose simply from our experience of the world. He notes: "The first thing impressed on us from our earliest infancy is, that events do not succeed one another at random, but with a certain degree of order, regularity, and connection…" (Stanley 2015, 34; Herschel 1851, sec. 26, 35). For Herschel, who was a strong supporter of a Baconian approach to science, the uniformity of nature was a prerequisite for being able to do inductive, experimental science, since without some uniformity one could not experimentally or inductively develop any laws of nature which "are not only permanent, but consistent, intelligible, and discoverable" (Herschel 1851, sec. 32, 43–44). As Stanley shows, the idea of nature being uniform and governed by laws would be seen by the end of the nineteenth century as a fundamental principle of secular, or naturalistic, science which denied any divine action in the world, but during the early Victorian period, the uniformity of nature was also a key principle of theistic science (Stanley 2015, 34–35). Even though Herschel's *Preliminary Discourse* might not be seen as overtly theistic, he does make a connection between laws of nature and God. He says:

> The Divine Author of the universe cannot be supposed to have laid down particular laws, enumerating all individual contingencies, which his materials have understood and obey; – this would be to attribute to him the imperfections of human legislation; – but rather, by creating them, endued with certain fixed qualities and powers, he has impressed them in their origin with the *spirit*, not the *letter*, of his law, and made all their subsequent combinations and relations inevitable consequences of this first impression, by which, however, we would no way be understood to deny the constant exercise of his direct power in maintaining the system of nature, or the ultimate emanation of every energy which material agents exert from his immediate will, acting in conformity with his own laws.
>
> (Herschel 1851, sec. 27, 37)

Stanley argues that one of the best examples of a theistic approach to science is Baden Powell's book *The Connexion of Natural and Divine Truth* (1838) (Stanley 2015, 35–37). Like Herschel, Powell argued that science would be impossible without some uniformity in nature but makes a more direct link between that uniformity and God. It is only through the uniformity and regularity of nature that humans can discover the physical causes that govern nature, and those physical causes serve as prerequisites for the discovery "of a regulating moral cause; and deduce the conclusion of a super-intending volition and designing intelligence" (Stanley 2015, 36). That is, science requires uniformity and the source of that uniformity is the existence of God.

The theistically rooted idea of a uniformity of nature led scientists to begin to search for some unity in nature through discovering some unification of scientific laws and discovering connections between different types of phenomena. One of the leaders in describing the connections between the sciences was Mary Somerville, whose book *On the Connexion of the Physical Sciences* (1834) influenced both scientists and nonscientists and as we have seen George Wilson recommended it to his future sister-in-law. Tellingly, on the title page, Somerville quotes Francis Bacon as saying: "No natural phenomenon can be adequately studied in itself alone – but to be understood, it must be considered as it stands connected with all of Nature." In the preface of her book, she outlines some of the connections that had recently been discovered. She says:

> The progress of modern science, especially within the last five years, has been remarkable for a tendency to simplify the laws of nature, and to unite detached branches by general principles. In some cases, identity has been proved where there appeared to be nothing in common, as in the electric and magnetic influences; in others, as that of light and heat, such analogies have been pointed out as to justify the expectation, that they will ultimately be referred to the same agent: and in all there exists such a bond of union, that proficiency cannot be attained in any one without a knowledge of others.
>
> (Somerville 1840, preface)

Somerville was clear that the root of the connections she was describing was the existence of God. After talking about the importance of the study of astronomy, and especially gravity, for an understanding of how everything in the universe, both planets and atoms, are connected, she goes on to say: "Equally conspicuous is the goodness of the great First Cause, in having endowed man with faculties, by which he can not only appreciate the magnificence of His works, but trace, with precision, the operation of His laws ..." (Somerville 1840, 2). Here she is drawing on both the idea that the laws of nature are rooted in God and also drawing on the Common Sense philosophy notion that God has created an intuitive element in the human mind that allows humans to discern the order that God has created. She then goes on to argue that Newton's law of gravitation was "one of those powers, which the Creator has ordained, that matter should reciprocally act upon matter" and goes on to argue that "gravitation must have been selected by Divine Wisdom out of an infinity of others, as being the most simple, and that which gives the greatest stability to the celestial motions" (Somerville 1840, 3, 430).

Unity

As we discussed in the first chapter during the first half of the nineteenth century, a number of scientists were searching for the interconnections between different types of natural phenomena. The result of this effort would be the formulation of the conservation of energy during the 1850s (Smith 1998). The concept of energy would provide a unifying factor connecting different types of natural phenomena. Thomas Kuhn argued that a number of discoveries concerning the interconnection of physical phenomena contributed to the formulation of the idea of the conservation of energy (Kuhn 1959 321–356). During the first half of the nineteenth century, there were discoveries concerning the relationship between heat and light, between heat and electricity, between heat and work, and between electricity and magnetism. Not everyone contributing to the development of the conservation of energy was practicing theistic science, but some were. In his biography of Michael Faraday, L. Pearce Williams argues that Faraday's Sandemanian faith played a role in his "proof" of the conservation of energy, what he called the conservation of force (Williams 1971, 4). Sandemanians were followers of the dissident Scottish preacher, Robert Sandeman, who, among other things, preached that the existence of God was expressed through the intricate patterns that existed in nature. In a lecture on ozone, Faraday referred to the "glimmerings we have … by which the *one Great Cause* works his wonders and governs the earth" (Williams 1971, 103). He was not making an argument from design that the designs in the world could prove the existence of God, rather he specifically rejected this idea when he said: "Let no one suppose for a moment that the self-education I am about to commend in respect of things of this life, extends to any considerations of the hope set before us, as if man by reasoning could find out God" (Williams 1971, 103). Instead, in the tradition of theistic science, Faraday began with the assumption that God existed and because of this, one could discover the intelligible and the beautiful in the world.

In a similar manner, Stanley argues that James Clerk Maxwell's evangelical faith impacted his science, especially his paper "On Physical Lines of Force," which helped to unify the phenomena of electricity and magnetism (Stanley 2015, 39). Maxwell was born in Scotland in 1831 and came of age during the period of the "Disruption" that split the Church of Scotland. He attended the church in St. Andrews where the Disruption began but was also schooled in the Episcopalian faith of his mother (Stanley 2015, chap. 1). He studied at Cambridge when natural theology was on the rise and the *Bridgewater Treatises* were being published. While reading Thomas Browne's *Religio Medici*, a seventeenth-century religious testament that drew heavily on scientific examples, and was a favorite book of George Wilson, Maxwell collapsed and emerged as a devout evangelical (Stanley 2015, 15–16). His new evangelicalism led Maxwell to argue for a theistic approach to the world, stating that "I think that each individual man should do all he can to impress his own mind with the extent, the order, and the unity of the universe …" (Stanley 2015, 37). Stanley argues that for Maxwell, the concept of unity was not simply a theological concept but equally a scientific concept and quotes him as saying: "We see that the hypothesis [of uniformity] coincides with all which Science and Religion alike teach, respecting the invariability of His mode of working" (Stanley 2015, 39). Stanley also argues that a powerful tool of the theistic approach to science was the use of analogy which Maxwell defines as the "partial similarity between the laws of one science and those of another which makes each of them illustrate the other" (Stanley 2015, 40). Maxwell would make good use of this approach when he developed a mechanical model of gears and idle wheels in order to explain electromagnetism.

Another scientist who may have influenced a theistic view of science was the Prussian naturalist, geographer, scientific traveler, and philosopher, Alexander von Humboldt. His multivolume work *Cosmos: A Sketch of a Physical Description of the Universe* (1845–1862) was one of the most influential works of the middle of the nineteenth century, selling out its first volume and then being translated into most major languages. In a letter to a young friend, dated January 15, 1847, Wilson specifically recommended that he read Humboldt's *Cosmos* (G. Wilson 1862d, 6). The book became one of the best-selling works of the mid-nineteenth century, and it became so influential that several years ago Susan Faye Cannon argued that instead of labeling early Victorian science as being Baconian, it should be relabeled as "Humboldtian science" (Cannon 1978, 73–110). For her Baconian science gave the impression of scientists or naturalists going into the field and simply collecting facts or objects without having any preconceived theory and then expecting that those facts will somehow produce general overarching laws through a vague application of an inductive method (Cannon 1978, 73). More recent work on Bacon would dispute this simplified version of Bacon's philosophy, and Cannon does argue that very few, if any, true naturalists were Baconian in how she defines it since the collection of objects and facts required some predetermined plan and categories in order to recognize something unusual that needs to be explained. More of a problem was that many of the scientists advocating a Baconian method were more often advocating a Newtonian method of analysis and deduction. But Common Sense philosophers saw no problem in reconciling Bacon and Newton. Still, Cannon argues that rather than using Bacon as a model for early Victorian science, Alexander von Humboldt would be more appropriate and advocates for the use of the term "Humboldtian science." But, as we will see, Bacon was still enormously influential among Scottish scientists and among Common Sense philosophers, at least through the mid-century. We will also see that Wilson himself was clearly influenced by Bacon's philosophy. It is true that by the nineteenth century through the writings of Dugald Stewart, many of Bacon's original ideas were undergoing revision, such as the role of hypotheses in the development of a scientific theory, and it must be noted that Bacon himself accepted the hypothesis of atomism. It might be better to see Humboldtian science not as a replacement for Baconian science but as an adjunct to it. Wilson certainly was able to accept both the scientific style of Humboldt and the style of Bacon without feeling the need to choose between them.

Humboldt was born in 1769 and had an early interest in collecting. While a student at Göttingen, he met Georg Forster, who had traveled with Captain

James Cook. The two traveled together around Europe and met with Joseph Banks, who had also traveled with Cook. In preparation for a career in science, Humboldt took some trips in Europe collecting mineralogical and botanical information and he enrolled at the Freiberg School of Mines. This resulted in a position in the Prussian Department of Mines. During the period, he became friends with Goethe. Humboldt had a passion for travel and in 1797, he resigned from his mining position in order to prepare to become, what he called, a "scientific traveler" (he never liked the term explorer). After gaining additional skills, and obtaining equipment and a boat, he left in 1799 for a five-year journey to South America, spending much of his time in what is now Venezuela, tracing out the Orinoco River, and traveling through the Andes in Peru. He also visited Cuba, what is now Mexico, and during his return journey stopped in Washington, D.C., and Philadelphia. After his return to Paris, he spent the next two decades publishing the results of his research in the Americas in a series of over thirty volumes (Dettelbach 1996, 288).

Humboldt's research was aimed at creating what he called *physique du monde* (terrestrial physics) that would use scientific instruments, such as thermometers, barometers, sextants, compasses, chronometers, and much more, to measure and map everything that varied with geography (Cannon 1978, 81). The goal of this new terrestrial physics was something close to the goals of both transcendental and theistic naturalism. At the very beginning of his book *Cosmos*, which became a summary of his research and his view of nature, he says:

> In considering the study of physical phenomena, not merely in its bearings on the material wants of life, but in its general influence on the intellectual advancement of mankind, we find its noblest and most important result to be a knowledge of the chain of connection, by which all natural forces are linked together, and made mutually dependent upon each other ... Nature considered 'rationally', that is to say, submitted to the process of thought, is a unity in the diversity of phenomena; a harmony blending together all created things, however dissimilar in form and attributes; one great whole animated by the breath of life. The most important result of a rational inquiry into nature is, therefore, to establish the unity and harmony of this stupendous mass of force and matter ...
> (Humboldt 1858, 1, 23–24)

This unity in diversity, which is very similar to a term Wilson uses, would then lead naturalists to discover "analogies between distant climates, vegetations, topographies, even cultures" (Dettelbach 1996, 299). For example, Humboldt argued for the importance of the relationship between plants and geography, especially how the distribution of plants was affected by altitude and climate (Cannon 1978, 83–84). He argued that a study of the geographical distribution of fossilized plants could provide information on the ancient geological connections between the continents or the changes in climate that have occurred over time.

While it probably cannot be claimed that Humboldt was practicing theistic naturalism, he does make a connection between nature and a higher spiritual world. He says:

> In order to trace to its primitive source the enjoyment derived from the exercise of thought, it is sufficient to cast a rapid glance on the earliest dawnings of the philosophy of nature, or the ancient doctrine of the 'Cosmos.' We find even among the most savage nations (as my own travels enable me to attest) a certain vague, terror-stricken sense of the all-powerful unity of natural forces, and of the existence of an invisible, spiritual essence manifested in these forces, which in unfolding the flower and maturing the fruit of the nutrient tree, in upheaving the soil of the forest, or in rending the clouds with the might of the storm. We may hear trace the revelation of a bond of union, linking together the visible world and that higher spiritual world which escapes the grasp of the senses. The two become unconsciously blended together, developing in the mind of man, as a simple product of ideal conception and independently of the aid of observation, the first germ of a 'Philosophy of Nature'.
>
> (Humboldt 1858, 1, 36–37)

Wilson's Unity in Variety

Beginning in 1845, shortly after he began his technical scientific researches, Wilson also began to lecture and publish on the non-technical aspects of science aimed at students and nonspecialists. Most of these originated as lectures and then were published. Much of this work attracted more attention than his technical scientific research (with the exception of his work on color blindness). These works covered a very wide range of subjects but can be seen to fall into three broad categories, although he tended to often work on all three simultaneously. One category involved lectures and publications that overtly addressed issues concerning the relationship of science and religion, such as natural theology or final causes in science, and might be seen as putting forth a theistic naturalism or theology of nature. A second category involved works on the history of science, particularly biographies of chemists and physicians, such as William Hyde Wollaston, Robert Boyle, John Dalton, John Reid, and Henry Cavendish. While his overtly religious writings seem to be aimed at arguing for an ideology of science based on a theology of nature, his histories of science seem more aimed at providing examples of a methodology of science based on Baconian philosophy and Common Sense philosophy. Finally, the third category involved purely popular subjects, such as chemistry and poetry, the electric telegraph, or the five senses. Even in these lectures and publications, there are usually some references to religion. One thing that all of his non-technical lectures and publications have in common is a focus on unity and the search for a unity behind the vast amount of variety that exists in the natural world. We will see in later chapters that the concept of unity also played a role in his work in technology. In his biography of William Hyde Wollaston, which was published in the *British Quarterly Review* in 1846, Wilson states: "Unity in variety is the voice of all nature" (G. Wilson 1862e, 274). But Wilson also saw unity and variety as having a dialectic relationship. In his paper "The Chemistry of the Stars," he labels one section "Unity in Variety" but goes on to argue that "Unity is in nature often nearest us exactly when variety seems of have put it furthest away." And then he goes on to say that like the sailors of Magellan who thought that they were traveling far away from home ultimately but ended up back where they started:

> We should set sail from Unity, and transverse the great circle of a uni-verse's variety till we came round to Unity again. The words on our lips as we dropt anchor would be 'There are differences of administrations, but the same Lord, and there are diversities of operations, but it is the same God which worketh all in all'.
>
> (G. Wilson 1859, 21)

While the concept of unity appears in almost all of Wilson's non-technical lectures and publications, it is particularly a focus of his works on science and religion. As we will see, these works drew from a wide range of sources, including transcendental naturalism, Common Sense philosophy, natural theology, a theology of nature, theistic science, and Humboldtian science, in order to first demonstrate that behind the great variety in nature there is an underlying design, archetype, or unity of organization. This belief in and search for some archetypal form or unity of organization clearly link Wilson with many of the transcendental naturalists, like Edward Forbes and Samuel Brown, who were Wilson's close friends and fellow members of the Brother-hood of Truth. But we will see that Wilson went a step further in attributing the unity in nature to the existence of God, finding in God the source of the universal archetypal forms (G. Wilson 1852b, 305). Here, we will see that he may have been drawing inspiration from his own new evangelical experience as well as the ideas of Thomas Chalmers and the Reverend John Cairns, whom he called his "spiritual father." Cairns, who had been a student of Sir William Hamilton, may have been important in providing a connec-tion between German transcendentalism, Common Sense philosophy, and evangelicalism.

Wilson's earliest popular writing was an essay entitled "On the Alleged Antagonism between Poetry and Chemistry," which was published in 1846 (G. Wilson 1846, 13–16). It was the concluding part of an introductory lec-ture on a course on the chemistry of gases given to an audience of women at the Edinburgh Philosophical Association on November 7, 1845. The lec-ture anticipates many of the themes that Wilson would address in his pop-ular writings, including the philosophy of Francis Bacon, Common Sense philosophy, and the role of God as a source of unity in the universe. The main purpose of the lecture was to convince his audience of women that a study of chemistry with all of its smoke, fumes, and acids would not "destroy a taste for other intellectual pursuits," especially an appreciation of beauty (G. Wilson 1846, 13). He begins by arguing that chemists are just as human as poets and the only conflict is between what he calls the "poetaster and the dabbler in science," but for true philosophers and poets, there is no conflict (G. Wilson 1846, 14). The reason that he sees for the lack of conflict is that God "created us with a love of knowledge for its own sake, of truth because it is true apart from any consideration of the beauty, terror, or other aesthetical or emotional quality which the truth may possess" (G. Wilson 1846, 14). Here, he quotes Bacon's reference to a proverb of King Solomon, saying "It is the glory of God to conceal a thing: but the honour of kings to search out a matter," and goes on to say "it has pleased God purposely to conceal from us, that our faculties may be exercised by penetration through the transparent veil, which hides them" (G. Wilson 1846, 14; Bacon 1937, 220). Here, Wilson seems to be echoing the Common Sense idea that the human mind does not simply reflect the external world through sense impressions but that the mind has an intuitive aspect that allows us to understand and engage the world. Also, the fact that Wilson attributes this intuitive aspect of the mind to God seems to echo Chalmers's theology of conscience.

This intuitive aspect of the mind allows Wilson to argue that poetry and chemistry may simply reflect different intuitive faculties of the mind. When we try to penetrate the veil and look into the causes of phenomena, "we play the part of philosophers, or students of science," but when we make use of the endowment that God has given us for a love of the beautiful, "we become for a time, though not perhaps professionally, poets, painters, sculptors, musicians ..." and "as all men are more or less philosophers, so all men are more or less poets" (G. Wilson 1846, 14). As such he claims that science and poetry differ from one another, but they are "not in opposition to each other," since: "The aim of science is truth. The desire of poetry in beauty, and in a glorious sense, all truth is beautiful, all beauty is true" (G. Wilson 1846, 15).

Wilson goes on to give examples of the role of chemistry influencing literature, such as the fact that Goethe taught Schiller the poetic capabilities of botany, or the fact that Coleridge, Southey, and Sir Walter Scott attended the lectures of Humphrey Davy, while Shelley conducted chemical experiments (G. Wilson 1846, 15). Wilson's conclusion anticipates another theme in his future writings. He states that science and poetry are like the binary stars recently discovered by John F.W. Herschel. Wilson says: "That star, or sun, for it is both, with its cold, clear, white light, is SCIENCE: that other with its gorgeous and ever-shifting hues and magnificent blaze is POETRY," and that they "shine towards that centre from which they came, even the throne of Him who is the source of all truth, and the cause of all beauty" (G. Wilson 1846, 16). Here, Wilson seems to be identifying science with unity because of its cold, clear, white light, what Bacon refers to as the *Lumen siccum* (Bacon

1884, 407; Bacon 1937, 178), and identifying poetry with variety because of its ever-shifting hues. As has been already argued, the idea of unity in variety would become a central element in Wilson's science and technology and we will see a number of examples later in the book. Finally, the fact that Wilson sees the source of this unity in variety as God would also come to be central to his development of a theology of nature.

Wilson's Evangelical Science

One of Wilson's earliest writings on science and religion was his paper entitled "Chemistry and Natural Theology" published in the *British Quarterly Review* in 1848 (G, Wilson 1862b, 1–50). The paper was written in response to William Prout's *Bridgewater Treatise* on "Chemistry, Meteorology, and the Function of Digestion, considered with Reference to Natural Theology," and George Fownes's Actonian Prize Essay on "Chemistry as Exemplifying the Wisdom and Beneficence of God." He states at the beginning that his purpose is to give "readers some conception of the way in which chemistry assists, as well as perplexes, natural theology" (G. Wilson 1862b, 2). While he says that one could argue that "chemical substances are regulated by laws most uniform, most simple, and harmonious, and proceed thereafter to infer that there must have been an Author...," he wants to instead argue that the proof for the existence of God provided by chemistry takes on significance only when it is discussed in terms of its role in providing living beings with the necessary elements to function and survive (G. Wilson 1862b, 2–3). As an example of some purposeful design, Wilson focuses on the chemical composition of the earth's atmosphere which he also sees as a unifying concept since "every living being on the dry land is bathed in it, and lives on it, and by it, and that those that are in the sea drink it in, dissolved in the element in which they live ..." (G. Wilson 1862b, 3). But not only does the atmosphere sustain animal life on land and in the sea, it also interconnects plants and animals allowing plants to provide food for animals and animals to provide food for plants. Wilson then goes through a detailed argument to show that out of a very large number of chemical gases, the ones that make up our atmosphere, such as oxygen, nitrogen, and others, are the perfectly designed ones, and in the perfect ratios, to sustain both plant and animal life. As such, he notes: "We must now try to conceive of the atmosphere as a whole, and to realize clearly the idea of its unity. And what a whole! what a unity it is!" (G. Wilson 1862b, 18).

Up to this point, Wilson seems to be making a typical argument from design, arguing that the composition of the atmosphere could not be the result of simply random forces but the action of some designer. But here, he raises his first problem with natural theology – that a study of nature can show the necessity of a designer but not of a single designer (G. Wilson 1862b, 23). He seems to solve this problem by resorting to Common Sense philosophy and arguing for the "appeal to the love of unity in every man's breast," and "that there is an *a priori* intuition in our minds of one God"

(G. Wilson 1862b, 23). But for Wilson, there is a much more troubling problem that cannot be so easily solved by natural theology. While natural theology can demonstrate the wisdom, knowledge, and power of God, he is not sure if it can show God's beneficence or goodness. He raises the problem that the same atmosphere that sustains animal and plant life can also bring winter rains, cold frosts, hailstorms, and spread diseases and plagues. He goes on to raise the issues of death, killing, pain, and suffering that are as prevalent in nature as is design. Given his extensive medical problems, Wilson was probably especially interested in the problem of pain and suffering. Every living thing is designed to die, and in the animal world killing another animal is essential to survival and this killing is not humane but often involves pain and suffering. He then goes on to say: "It is the absence of anything like a resolute attempt to look this great problem of physical evil in the face, that renders our *Bridgewater Treatises* so little valuable as works on natural theology." But as noted earlier, he goes on to say: "We except entirely from this charge Dr. Chalmers' beautiful volume..." (G. Wilson 1862b, 20).

Wilson's praise for Chalmers's *Bridgewater Treatise* can give us some insights into Wilson's ideas concerning natural theology and chemistry. Beyond the *Bridgewater Treatise*, Wilson may have been also influenced by Chalmers's theology given the fact that W. Lindsay Alexander, Wilson's minister, had been a pupil of Chalmers at the University of St. Andrews. Also, as noted earlier, Chalmers had attended Wilson's lecture on transmutation in 1844. Like Chalmers, Wilson is critical of the standard natural theologies for not even trying to address the problem of evil saying:

> It seems to us, therefore, a plain and imperative duty to illustrate ... the extent to which chemistry reveals evil as well as good in the world, and thereafter to consider ... how far the existence of that evil modifies our views of the benevolence of God.
>
> (G. Wilson 1862b, 27)

Wilson presents two examples that lead to questions about God's goodness. His first example is that the fossil record seems to indicate that the climate in the northern latitudes was at one point much warmer than the present and the change in that climate resulted in the mass destruction of animal life in those regions (G. Wilson 1862b, 33). His second example is the shark that clearly exhibits purposeful design, but that design is aimed at killing (G. Wilson 1862b, 42–43). This seems very similar to Chalmers' argument that an instrument of torture can exhibit the idea of design. Here, both Wilson and Chalmers seem to be doing what Boyd Hilton has called turning natural theology on its head (Brooke 1999, 26). That is, evangelicals often used the search for design to discover in nature chaos, extinctions, and a ruined world. Wilson's solution to this problem also seems to have possibly been influenced by Chalmers's idea of a theology of conscience or what Topham calls a natural theology of conscience. In talking about the problem of not confronting the problem of evil, Wilson says: "It is in treatises on the physical sciences that the defect we are lamenting is most liable to occur; for psychology and ethics cannot possibly be discussed without compelling the consideration of evil as well as good" (G. Wilson 1862b, 47). Also like Chalmers, Wilson argues that the debate concerning God's goodness can only be solved by focusing on the intuitive aspects of the mind, what Chalmers labels as conscience but Wilson calls the heart. Previously, in his essay on poetry and chemistry, Wilson had noted that the mind had two faculties, facts that were apprehended by the intellect and truths that set the heart on fire (G. Wilson 1846, 14–15). Wilson says:

> So long as men look upon the origin, and existence of moral and physical evil as a problem which can be solved by logic, they will struggle to the very death to reach the solution; but when they discover that in this world a solution of the difficulty cannot be attained, they will cease to combat with it, and transfer it from the region of the intellect to that of the heart, as a sad and solemn mystery which, with closed lips, will haunt them to their graves
>
> (G. Wilson 1862b, 49)

This reference to the problem of evil as a mystery also seems to reflect Chalmers' argument that natural theology can raise questions, but it cannot answer those questions.

Ultimately Wilson's solution to the problem of evil seems to draw more from Topham's idea of a theology of nature that combines natural theology with revelation. Wilson argues:

> if an accredited and trustworthy divine relation shall have assured us of the unity of him who has declared that 'the Lord our God is one Lord,' then physical science will affirm that all creation entirely accords with such a declaration.
>
> (G. Wilson 1862b, 23)

That is, for Wilson, the solution to the problem of evil is rooted in the concept of unity. He argues that the "evil and the good in nature are inextricably intertwined, and cannot be unraveled or disentangled from each other" (G. Wilson 1862b, 48). That unity may be found in the physical world through natural theology, but the source of that unity is found through revelation. He concludes: "even in this world, all who believe in revelation may contemplate with a joyous eye the good that is in it, and adjourn the explanation of evil as something traversing, but not neutralizing or annihilating its opposite" (G. Wilson 1862b, 50). That is, what Wilson finds most important is an argument *for* design in the world based on the existence and unity of God rather than an argument *from* design.

In an address to new medical students, Wilson again expresses some criticism of traditional natural theology. In his "Introductory Address Delivered at the Opening of the Medical School, Surgeons' Hall, Edinburgh" on November 6, 1850 (G. Wilson 1850b), Wilson states that the purpose of his address is "a consideration of the temper or spirit in which the study of medicine should be entered on and prosecuted" (G. Wilson 1850b, 3). One of his main arguments is that a student of medicine must be trained "to the law as to the fact, to the theory as to the practice, to the reason as to the rule," and in order to do this medicine must educate "simultaneously the intellect and the senses" (G. Wilson 1850b, 5). This can be seen as a recurring theme in Wilson's writings. In a later chapter, we will see that Wilson suggested that the symbol for the Industrial Museum of Scotland be an eye in the palm of an open hand, combining the intellect with the senses. This combination of the intellect with the senses might also reflect Common Sense philosophy which relied heavily on the senses but argued that those sense impressions had to be interpreted through some innate qualities or powers of the mind. In fact, Wilson makes a specific point that medical students cannot rely on their senses but that their senses require training and need to be educated (G. Wilson 1850b, 8). Wilson goes on to warn students about "false science" that seems to be a reference to the biblical category of "science falsely so called" (1 Timothy 6: 20). As a way to avoid false science, Wilson uses the example of Francis Bacon's "Idols" and presents the goal to be able to use what Bacon called the "*Lumen siccum*," or the "pure un-tinted light of reason, which allows truth to be seen without giving it a colour" (G. Wilson 1850b, 10). Again, in order to accomplish this, Wilson argues for educating our senses so we become honest observers.

At the end of his lecture, Wilson turns to moral and religious issues by arguing that "Science, gentlemen, is another word for the works of God" (G. Wilson 1850b, 12). He warns that by its nature, medicine's focus on the human body can lead to a belief in materialism and even atheism. In what seems to be a call for final causes, he notes that the focus on secondary causes "are apt to make us forget that what we call laws and forces – physical, chemical, physiological – are but the modes of working of Him who is uncaused, and the author of director of all" (G. Wilson 1850b, 13). But, Wilson also goes on to admit that when confronted with disease, blindness, deafness, and madness in the hospital, it seems that design fails and he calls for a chapter to be added to the *Bridgewater Treatises* in order to reconcile these conditions with design. He admits that a diseased body can still be evidence of design, but he says: "It is the dark, not the bright side of the argument from design for an all-wise and all-merciful Creator, which demands elucidation at the hands of our natural theologians, but it seldom receives it" (G. Wilson 1850b, 14). Here, Wilson seems to have "turned natural theology on its head."

Another work in which Wilson addressed the relationship between science, and especially chemistry, and religion was his 1852-paper "The Chemistry of the Stars: An argument Touching the Stars and Their Inhabitants," which he oddly published as part of a book on *Electricity and the Electric Telegraph* (G. Wilson 1862c, 51–103; G. Wilson 1859). One stimulus for writing the work was the publication of a new edition of John F.W. Herschel's *Outlines of Astronomy* (1849) that presented a number of new astronomical discoveries based on new larger telescopes (G. Wilson 1862c, 51–52). In the book, Herschel hints at the possibility that the moon might be inhabited, although not necessarily with earth-like creatures (Herschel 1849, 261–262). Debates concerning the existence of life on other planets, or the plurality of worlds, have a long history going as far back as the Epicureans, and given support by Giordano Bruno, the Reverend John Wilkins, Newton, Leibniz, Thomas Chalmers, and others (Crowe 1986, chap. 1). Herschel's father William was a firm believer that life existed on all of the planets of the solar system and beyond. Although John Herschel's reference to lunar inhabitants in his *Outlines* is brief, in reports made from the Cape of Good Hope in 1835, he claims to have observed animal-like creatures with his new telescope (Crowe 1986, 210–211). Wilson may have been especially drawn to Herschel's argument that the existence of other inhabited worlds was part of God's overall plan (Crowe 1986, 217). After asking what is the purpose of the vast number of stars and planets, Herschel says: "who can suppose man to be the only object of his Creator's care, or who does not see in the vast and wonderful apparatus around us provision for other races of animated beings" (Herschel 1849, 554).

While there was a spirited debate during the mid-nineteenth century over the possibility of life on other worlds, there was a further debate over what that life might look like – would it be the same as life on the earth or would it be dramatically different. Wilson is somewhat vague concerning whether other planets contain life. He says: "We leave, then the question of the universal habitation of the heavenly bodies untouched, and intend, moreover, to refer chiefly to the nature of the stars, and not to that of their inhabitants" (G. Wilson 1859, 5). In his preface, he seems to leave the door open to life on other planets but concludes that: "The unequivocal testimony, then, of physical science, as it seems to us, is against the doctrine that life, as it appears on the stars, must be terrestrial in its nature ..." (G. Wilson 1859, vi). Rather than discussing the possibility or nature of extraterrestrial life, Wilson's main purpose is to argue that the physics and chemistry of the planets and the stars are dramatically different than the earth's and "to claim for our earth uniqueness as an abode of living creatures," or to show that the probability is against other worlds like ours (G. Wilson 1859, vi–vii).

It turns out that Wilson's ultimate purpose was not to enter into a debate about the plurality of worlds but to put forward a critique of *Vestiges of the Natural History of Creation*. The work was anonymously published in 1844 but was later discovered by the work of the Edinburgh publisher and naturalist, Robert Chambers. As James Secord shows the work immediately became a sensation in mid-Victorian Great Britain (Secord 2000). The book put forward an evolutionary view of the creation of the solar system and of human life. It argued for the nebular hypothesis in which the planets of our solar system formed from a giant whirling cloud of dust, and it went on to discuss the progressive development of animals from fish to amphibians to reptiles and to mammals and suggested that humans might have evolved from lower animals, such as apes. The work caused immediate religious debates, just as Charles Darwin's work would do a few years later. This debate was amplified in Scotland since the book came just after the Disruption of 1843 which led to a split in the Church of Scotland Secord 2000, chap. 8). The book also became controversial because of its association with phrenology that was particularly popular in Edinburgh at this time (as it turned out, the book's author was a supporter of phrenology). George Combe, one of the founders of phrenology in Edinburgh, drawing on the writings of Franz Joseph Gall, believed that a science of the mind had to precede any understanding of natural science (Secord 2000, 69–76). Phrenology was based on the belief that different mental functions were centered in different parts of the brain and these could be strengthened through different types of mental "exercises" but had to be kept in some balance. Some of phrenology's ideas came into conflict with evangelicals (Secord 2000, 271–274). Phrenology placed emphasis on the physical act of reading but argued that reading and understanding the Bible could only occur after the various faculties of the mind had been exercised through reading other works, such as natural philosophy. This seemed like a model for natural theology in which one first studied nature in order to appreciate the mind of God, but it was also the exact opposite of the approach of many evangelicals. In their theology of nature, revelation, and therefore the Bible, had to come first and only then could one understand the designs in nature.

The *Vestiges* association with phrenology and with a deist-like progressive theory of nature led to some immediate attacks. Secord notes that the "central issue was God's constant watchfulness over the creation and his ability to act in any way," and that "the inexorable upward progress of the development hypothesis denied the need for the miracle of Christ's atonement ..." (Secord 2000, 281–282). In Scotland, Wilson was one of the leading critics of the *Vestiges*. Secord notes that Wilson wrote an anonymous critical review for the *British Quarterly* in 1845 and possibly a longer review for the same journal a year later, and Secord also notes that Wilson gave a series of four lectures criticizing the science in *Vestiges* to the Young Men's Society, but they were unfortunately never published even though several publishers were interested in doing so (Secord 2000, 285–286).

In this light, it appears that Wilson saw his *Chemistry of the Stars* as a response to *Vestiges*, which he mentions several times in his work. Crowe notes that near the end of *Vestiges* that evolutionary theory does not just apply to the earth but to the "apparently infinite globe-peopled space," a conclusion that Chambers based on his belief that "all planets are so similar to the earth that the inhabitants of all other globes of space bear not only a general, but a particular resemblance to those of our own" (Crowe 1986, 224–225). Wilson's fundamental objection to *Vestiges* was the author's "blind zeal for the nebular hypothesis of a common physical origin of all worlds," which then becomes "solicitous to save God the trouble of taking care of his own universe"

(G. Wilson 1859, 43). This leads him to try to show that the chemistry of the stars and planets differs from the earth's chemistry. The first part of his argument is to use a "commonsense approach" by bringing together a "jury" of average people, such as a chandler, a blacksmith, a sailor, a soldier, and a gentleman. He provides them with the newest forty-foot telescope to examine the planets and the heavens and asks them to decide if all of the stars and planets look the same (G. Wilson 1859, 7–21). Each "juror" using their own experience in their own profession concludes that there is so much diversity in the heavens that the stars and planets cannot all be the same. For example, the sailor compares the stars and planets to ships sailing through the heavens but notes from his own experience how many different ships exist on just our planet. He ends up concluding that: "If there are all of these differences among our sailing vessels, are there likely to be fewer among the ships of heaven?" (G. Wilson 1859, 14). The others make similar conclusions. If the stars are an army, there is great diversity in an army with cavalry, infantry, artillery battalions, rocket companies, engineers, and sappers. If the stars are like lamps, the earth has candles, torches, gas lights, electric lights streetlights, and lighthouses. The foreman of the jury renders a verdict by quoting *I Corinthians* 15:41: "There are celestial bodies, and bodies terrestrial: but the glory of the celestial is one, and the glory of the terrestrial is another" (G. Wilson 1859, 20). While the verdict seems to confirm the diversity in the universe, Wilson warns against going too far and continues his argument for unity in variety. As earlier quoted Wilson sees a dialectic relationship between unity and variety saying: "Unity is in nature often nearest exactly when variety seems to have put it furthest away" (G. Wilson 1859, 21).

Wilson then turns to the opinion of who he calls the priestly dignitaries – the astronomer, the chemist, and the physiologist or biologist. It is here that he finds the most criticism of *Vestiges*, arguing that the actual science is wrong when it comes to the idea of the chemistry of the stars being the same as the chemistry of the earth. Using some of the ideas of John Herschel, Wilson argues that from an astronomical point of view, the planet and the stars are likely to have a more non-terrestrial character than a terrestrial one (G. Wilson 1859, 22–25). He bases this on the discoveries in astronomy that the planets receive differing amounts of heat and light from the sun, that the gravitational forces would differ in intensity, and that the nature of their materials would differ from the earth.

Not surprisingly most of Wilson's scientific criticism of *Vestiges* centers on chemistry (G. Wilson 1859, 26–40). Since the foundation of *Vestiges* was the nebular hypothesis, an important conclusion was that "inorganic matter must be presumed to be everywhere the same" and that other worlds are constructed out of the same material as our world. In support of this idea, *Vestiges* argues that meteorites contain the same chemicals as found on the earth. Wilson accepts that the stars and planets do not seem to contain new chemicals not found on the earth, but he challenges the idea that meteorites are earth-like by noting that "they contain only *some* of them" (G. Wilson 1859,

29). He notes that meteors contain only about one-third of the elements known in his time and that most of them are metals, while he argues that the earth's crust is composed of large amounts of oxygen that is necessary for terrestrial life and that most of the elements missing from meteorites are the chemicals most abundant in plants and animals. On the other hand, Wilson notes that a number of elements, such as selenium, vanadium, and others, are quite rare on the earth but may play an important role in non-terrestrial chemistry. The fact that meteorites and the earth differ in the variety of elements, the relative quantities of elements and the condition of those elements led Wilson to argue against the idea that the non-terrestrial world is chemically similar to the terrestrial one.

Finally, Wilson turns to physiology or biology (G. Wilson 1859, 40–46). The author of *Vestiges* argues that "Where there is light there will be eyes; and these, in other spheres, will be the same in all respects as the eyes of tellurian animals ..." (G. Wilson 1859, 44). Wilson responds that the differences in heat, light, and gravity noted by astronomers would make it difficult for earth-like life to exist on other planets as would the lack of oxygen on those planets. Wilson still leaves open the possibility that some form of life might exist on other planets, but such life would be significantly different from the life on our planet. While different worlds might be different from the earth, Wilson concludes with the idea that in this scientific diversity there is still a unity based on religion. He ends with a Biblical reference to *John* 14:2 and says: "In our Father's house are many mansions, and the Great Shepherd watches over countless flocks, and has other sheep which are not of this fold" (G. Wilson 1859, 50).

Shortly after, Wilson published his *Chemistry of the Stars*, an anonymous book appeared in 1853, entitled *Of the Plurality of Worlds: An Essay* that went beyond Wilson and claimed that no life exists on other planets (Whewell 1853). The author was later to be discovered as William Whewell, who ironically in his earlier *Bridgewater Treatise* of 1833 had supported the idea of a plurality of worlds. Michael Crowe notes that John Hedley Brooke argues that Whewell's change of mind was in response to the publication of the *Vestiges*, especially its evolutionary implications, and Crowe himself argues that Whewell saw that pluralism could not be reconciled with Christianity (Crowe 1986, 266–267). Whewell's *Essay* led to a long bitter debate with Sir David Brewster who advocated strongly for the plurality of words (Crowe 1986, chap. 7). Wilson, not knowing the identity of the author of the *Essay*, received a copy from Whewell and had praised the work in a latter edition of his own book, noting that the *Essay* gave additional support to his book since it made the argument from the point of view of astronomy and geology rather than chemistry (Crowe 1986, 327–329; G. Wilson 1859, v-vii). In response, Whewell in the preface to the third edition of his book noted that

> Dr George Wilson also has, in his lively tract on 'the Chemistry of the Stars,' made some very ingenious reflexions, tending to shew that the

earth, the planets, the stars and the sun, are probably very different from one another.

(Whewell 1855, 4)

During 1856, Wilson wrote two papers dealing with natural theology and the argument from design. The first paper was entitled "On the Character of God as Inferred from the Study of Human Anatomy" and was an address to medical students sponsored by the Edinburgh Medical Missionary Society (G. Wilson 1856, 25–99). In this paper, Wilson provides evidence both for and against natural theology and the argument for design. He begins by arguing that the observation that bodies exhibit what could be called instruments or apparatus fitted to produce certain effects and, using some of the ideas of Common Sense philosophy, argues that a belief that such designs are "created for a wise end" is something that we inherently believe because it has been "indelibly printed on our hearts" (G. Wilson 1856, 29). As proof that the world has been designed for human life, Wilson anticipates the modern anthropic principle and argues that if such a design did not exist, we would not be here to ask about the conditions of human life (G. Wilson 1856, 33). He goes on to conclude that the bodily instruments that living creatures possess lead to the conclusion "that the Maker and Maintainer of those wonderous living machines must be God" (G. Wilson 1856, 33). But after arguing for the belief in a personal designer, he goes on to argue that contradictions or qualifications to that belief must be considered. Using examples from his own medical training, Wilson raises some things that might provide obstacles to believing in divine power, wisdom, and goodness. For example, the condition of the corpses in the dissecting rooms, or treating patients in the hospitals, or facing the issue of disease could become a hindrance to a belief in apparent design (G. Wilson 1856, 38–50). Here, Wilson follows Chalmers's criticism of natural theology that we have already discussed and says: "let us not stand up to affirm to others that Good is Good, and refuse to listen when they reply that Evil is Evil" (G. Wilson 1856, 51). Wilson answers this problem by arguing that no design is perfect and that even Paley's watch would have not been perfect and would have not kept perfect time, but it still exhibited design (G. Wilson 1856, 55–62). He goes on to argue that such imperfections make examples of design stand out all the more. So that even imperfect, human hands and eyes can make surgical instruments, microscopes, steam engines, railways, and telegraphs.

Wilson then goes on to discuss that even if the idea of beneficent design cannot be logically disproved, there might be examples that make it difficult to believe in it. His first example is that not all organs of an animal are useful or purposeful. In answering this, Wilson makes a significant contribution to the problem of reconciling the doctrine of organization, or morphology, with the doctrine of final causes, or teleology (G. Wilson 1856, 63–75). Wilson most likely had been introduced to the morphological idea that there existed some higher level unity of plan or unity of organization that connects organisms through his association with the transcendental naturalists who were part of Edward Forbes's circle and his Brotherhood. He may have also been drawn to the idea of the final cause through his reading of Whewell's *History of the Inductive Sciences* and his *Philosophy of the Inductive Sciences*, as well as Chalmers's *Bridgewater Treatise*, where they both refer to final cause numerous times (Whewell 1845; Whewell 1840; Whewell 1837; Chalmers 1835, 2: 42, 73, 206). Wilson also notes that both Richard Owen and John Goodsir, two leading morphologists, saw the "fullest harmony between morphology and teleology" (G. Wilson 1856, 65). Finally, Wilson may have been influenced by Peter Roget's *Bridgewater Treatise*, which provided an argument that God is able to combine "perfect efficiency of function with simple beauty of form" (Rehbock 1983, 57).

By drawing upon morphology, Wilson is able to explain the fact that not all organs of an animal are useful or purposeful, which seems an apparent violation of teleology or final cause. Since morphology is looking for a general law of organization, or unity of plan, an organ in individual animals may not have any utility but it can be connected to some "mighty plan" that governs "all creatures" (G. Wilson 1856, 65–70). As an example, Wilson discusses the bones in the flipper of a seal, which resemble bones of a human hand, a lion's paw, or a bird's wing, but have no use for grasping, killing prey, or flying. But such seal bones might have had some utility in the past or might have some utility at a future time. Therefore, utility, purpose, or final cause cannot be thought of in terms of an individual organism but must be thought of in terms of the vast variety of organisms. In doing so, design takes on a new interpretation. Wilson says: "If the same bones, by a small change, can become hand or paw, or fin or wing, is not the proof of special design all the stronger ..." (G. Wilson 1856, 74). The relationship between morphology and teleology also provides an example of one of Wilson's favorite themes – unity in variety. The fact that a hand, a paw, a flipper, and a wing can arise from a single archetypal plan provides evidence of unity in variety. He concludes that

> the homologies and analogies which link creature with creature, and make man and the lion, the eagle and the swordfish lay hold, as it were, of each other's hands (or quasi-hands), as in one sense children of a common father, furnish proof of the unity of nature of the universal Creator, such as no isolated study of single creatures could ever supply.
>
> (G. Wilson 1856, 75)

After arguing that all organs of an animal might not be useful to that animal but can still represent purposeful design, Wilson raises a number of other issues that might not logically disprove beneficial design but can hinder full belief in it (G. Wilson 1856, 75–86). For example, Wilson argues that not all organs or functions of animals are graceful or becoming, specifically referring to issues of digestion. Similarly, he argues that there are organs or functions of our bodies that are hurtful and painful to us, citing the example of childbirth. Using the example of carnivores, he argues that animals have endowments to inflict pain and suffering. Finally, he argues that while our lungs are designed to inhale air and our stomachs to digest food, our bodies are not designed to resist injurious aspects arising in the inorganic world, such as the dust from volcanoes, the effects of storms, or the effects of climate change.

In summary, Wilson concludes that the study of human anatomy leads us to "acknowledge the clearest proofs of power, wisdom, skill, vast purpose and great mercy, as belonging to Him who has made, and who sustains all things." But he also

> cannot deny that our bodies seem to us, in some respects imperfect; that they have strange marks of humiliation upon them; that even in health they are in certain circumstances tortured by agony, and that they are the subjects of endless painful diseases.
>
> (G. Wilson 1856, 86)

He sees that good and evil are always tied together and cannot be separated but are part of one system. As in his earlier writings where he seems to be following the ideas of Chalmers and other evangelical naturalists, he sees the conflict between good and evil in nature as a mystery that can only be understood by combining natural theology with revelation. He argues that both the Book of Nature and the Book of Revelation not only proclaim beneficial design but also seem to sanction suffering. They are related but as mirror images. He says:

> The Book of Nature is a joyful, sorrowful epic, beginning in the far past eternity with the happy birth of a new world, and closing with the advent upon earth of man, the last, the most gifted and the most miserable of its inhabitants. The Book of Revelation is a sorrowful, joyful epic, beginning with the ruin of man's innocence, and with the earth and all its creatures cursed for his sake, and closing with the farewell of the Son of God, as he returned to heaven with the promise that he should come again to make all things new.
>
> (G. Wilson 1856, 91–92)

Wilson is ultimately calling for an evangelical theology of nature in which purposeful design and the final cause will not result in the proof of the existence of God, rather the "image of the earthly will be fully understood, only when it has changed into the image of the heavenly ..." (G. Wilson 1856, 97–98). That is, revelation will provide the key to earthly morphology and teleology, not the other way around.

Near the end of 1856, Wilson gave a lecture entitled "Chemical Final Causes: As Illustrated by the Presence of Phosphorous, Nitrogen, and Iron in the Higher Sentient Organisms" to the Royal College of Surgeons (G. Wilson

1862a, 104–164). The main point of the lecture is to answer the question of why out of sixty-some known chemical elements, only some seventeen are seen in significant amounts in the human body and the lecture focuses on the role of phosphorous, nitrogen, and iron in human physiology (G. Wilson 1862a, 104–108). Based on the fact that only a limited number of chemical elements are found in the human body, Wilson uses the transcendental argument that nature is based on some higher, or archetypal plan. As we have seen, Wilson was probably influenced by transcendental naturalism through his friendship with Edward Forbes, John Goodsir, and other members of the Brotherhood of Truth and we have just seen elements of this approach in his paper on the "Character of God," written earlier in 1856. With the exception of Samuel Brown, another friend of Wilson and a member of the Brotherhood, most of those drawn to transcendental naturalism were anatomists, physiologists, and paleontologists but in this lecture, Wilson seems to apply it to chemistry and may have been influenced by Brown's transcendental chemistry. He argues that "no creature is a fortuitous concourse of atoms," but each "is as definite and constant in its chemical composition as it is in its mechanical structure, or its external form,... so that each plant and animal has a chemical as well as an anatomical individuality" (G. Wilson 1862a, 114).

Wilson then goes on to analyze what types of chemicals would have the qualities needed to function in a human body. He argues that in order to function, the human body needs chemicals that exhibit "great stability and great mobility," or in other words, "[e]verlasting change, and yet fixity" (G. Wilson 1862a, 117–118). This seems somewhat similar to Wilson's idea that nature exhibits unity in variety. But he sees the change that is happening in the human body as a specific type of change. He notes that solids and liquids, like blood and flesh, are continually changing into one another and both liquids and solids are continually changing into gases. He further argues that "the human organism is the continual subject of swift changes of its composition in opposite directions" (G. Wilson 1862a, 120). As an example, he cites the fact that one-half of the blood in the human body which is in the arteries is in one chemical state, while the other half which is in the veins is in a different chemical state.

Although Wilson only uses the term polarity once in his paper, his argument that change in the human body takes place in opposite directions seems to reflect this concept. The concept of polarity was a fundamental element of *naturphilosophie* which served as the basis of much of transcendental naturalism and shows up in the writings of Edward Forbes, William Whewell, and Richard Owen (Rehbock 1983, 11–26, 69, 81–108). Wilson's friend Forbes had referred to the idea of polarity since the 1840s, when he used it to explain and classify British fossils (Rehbock 1983, 103–113). To explain a class of fossils called cystoidea, he argued that members of the class were in a struggle between "progression toward a higher type" and a "negative or vegetative polar influence" (Rehbock 1983, 103). After a skeptical response to his ideas, he seemed to move away from the idea of polarity but beginning in 1854, just before becoming Regius Professor of Natural History at the University of Edinburgh, Forbes returned to the idea of polarity in a presidential address to the Geological Society and later to in a lecture at the Royal Institution, and the idea was the subject of one of his last lectures at Edinburgh (Rehbock 1983, 104). In these lectures, Forbes used polarity to explain the distribution of fossils throughout geological time. Traditionally, geologists divided the period of stratified rocks into the Palaeozoic, the Mesozoic, and the Cainozoic, but Forbes argued that the last two were in fact one period that he labeled Neozoic (G. Wilson and Geikie, 1861, 543). Looked at in this way, he noticed that the variety of fossils did not follow a pattern of linear progression but the "maximum development of generic types during the Palaeozoic period was during its earliest epochs; that during the Neozoic period towards its later epochs," and he claimed that the "relation between them is one of contrast and opposition, – in natural history language, is the relation of POLARITY" (G. Wilson and Geikie 1861, 544–545).

Rehbock argues that Forbes saw polarity as a divine idea that arose from an interaction between a "Creative Power" and "suitable conditions" or environmental forces (Rehbock 1983, 111).

As noted, Wilson used the word polarity only once in his paper, but the fact he and Forbes were lifelong friends and Forbes's work on polarity was taking place just before and just after his move to Edinburgh, it could be speculated that Wilson's paper, written just two years after Forbes's death, was drawing on ideas of polarity. At the beginning of his paper, Wilson contrasts the role of nitrogen and phosphorous and uses language that suggests polarity. He says: "the fiery phosphorous and the negative nitrogen are the two elements which, by their greater abundance in animals, and the part which they play there, most strikingly distinguish animals from plants" (G. Wilson 1862a, 123). Most of Wilson's paper is a detailed and often technical study of how phosphorous, nitrogen, and iron are particularly suitable as chemicals to be found in the human body. In all three chemicals, what he sees as important to their role in the human body is their chemical qualities of mobility and stability. He notes that phosphorous has the ability be susceptible to great many chemical changes but "it can change this mobile, restless, agonistic condition for one of passive indifference and great stability" (G. Wilson 1862a, 125). He notes that nitrogen is "at once very fixed and very variable in its properties," and that "nitrogenous bodies as a class, belongs as a distinctive property to utmost readiness to undergo change, and exactly because they contain an element indifferent to change" (G. Wilson 1862a, 144–146). Although the role of iron in the human body was not well understood at the time, Wilson notes that iron has the intermediate property between metals that are very oxidable and very unoxidable and cojoins mobility and stability (G. Wilson 1862a, 150–152). In addition, he speculates that iron's magnetic properties play some role in the body, especially in the nerves whose "peculiar force or polarity" resembles the electrical and magnetic force (G. Wilson 1862a, 156). Again, while this is his only use of the term polarity in the paper, it might be argued that all of the examples he gives of how phosphorous, nitrogen, and iron all chemically have the qualities of mobility and stability also reflect the idea of polarity. One might also argue that stability and mobility are also examples of unity and variety, another of Wilson's favorite themes.

Wilson's examples of how phosphorous, nitrogen, and iron have special chemical qualities of mobility and stability that make them of particular importance in the human body could be seen as providing an argument in support of final causes or an argument for design. But, as we have seen in his previous papers, Wilson is often reluctant to make a strong argument in favor of natural theology and instead seems to favor a theology of nature that combines natural theology with revelation and he seems to take this approach in the conclusion of this paper. He says: "I refer to the truths (in so far as they are truths) expounded in the preceding pages as *illustrations*, not *demonstrations* of final causes" (G. Wilson 1862a, 160). This could be interpreted as not claiming to have proved the existence of final causes but simply providing examples of them. Similar to his paper on natural theology, he may be arguing that final cause cannot be used to prove the existence of God, but that God is the source of final causes. Wilson admits that the doctrine of final causes is in "disrepute" among many scientists and that the search for final causes should not be the "chief object of scientific inquiry" but

> when guided by the *Lumen Siccum* … and then find our hearts swelling with rapture at the wondrous example which it affords of God's wisdom and power, we are traitors to ourselves and to our Maker if we refuse adoration.
>
> (G. Wilson 1862a, 160)

Lumen sixxum (dry or perfect light) was a concept used by Bacon to refer to an organizing or instinctive principle that exists before we have any sense experience of nature (Bacon 1937, 178). It could often be connected to the idea of the eternal and was an important element of Coleridge's philosophy that had influenced the transcendental naturalists (Richards, 2002). So, like his argument about design, Wilson is arguing that the final cause cannot be understood without drawing on some revelation arising from God. He says: "What we call a final cause, is not God's final cause, but only that small corner of it which we can comprehend in our widest glance" (G. Wilson 1862a, 162). Wilson argues that the disrepute associated with the final cause arises from mistaking the small corner for God's final cause. He ends the paper by arguing that Bacon's declaration that final causes were sterile like nuns of vestal virgins has been misunderstood and taken out of context (G. Wilson 1862a, 163). For Wilson final causes are sterile because they have not practical application and they have no practical application because, like the vestal virgins, they belong to God and provide "the most perfect of earthly witnesses to the being and perfection of God" so that "we must seek after, and love final causes" (G. Wilson 1862a, 163–164).

Wilson's religious writings provide some new insights into the relationship between science and religion in early Victorian Britain. First, religion was still an important aspect of much of the science of the time and of many of the scientists. Some of the leading scientists, like William Whewell, William Buckland, Charles Bell, and Peter Mark Roget, among others, were authoring *Bridgewater Treatises* on natural theology. Natural theology provided a number of benefits to early nineteenth-century scientists such as limiting excess theorizing by providing reasonable solutions that were agreeable to a number of scientists, providing a way to reconcile the mechanical philosophy with a religious point of view, and shielded scientists from accusations of radicalism by aligning science with a religious tradition (Livingstone, Hart and Noll 1999, 8). Wilson may have drawn inspiration from Roget's *Bridgewater Treatise* in his attempt to expand traditional natural theology's emphasis on final cause with the transcendental idea of unity of plan. In doing so, Wilson provides a way to incorporate transcendental naturalism into a theological framework. In his *Life of Dr. John Reid*, as part of a discussion of Richard Owen's transcendental anatomy, Wilson says:

> [W]e have only to add to the conception of a Universal Archetypal form the datum of the recognition of God as its author and applier, to find in it an argument in favour of the greatness, and knowledge, and wisdom of the Creator, such as the older physical natural theologies could not supply. It furnishes in particular a proof of the *oneness of counsel* of the Architect of the universe, and a probability therefore of their *oneness of being*, or the most striking and convincing kind.
>
> (G. Wilson 1852, 305).

But evangelical scientists, like Wilson, also saw weaknesses in the traditional natural theology. They often saw the natural theology based on Paley as too optimistic since it failed to deal with the inherent evil in the world and the fallen origin of human nature (Brooke 1999, 26). In response, Wilson, following in the ideas of Chalmers, develops a theology of nature which "turns natural theology on its head" (Brooke 1999, 26). Where along with finding beneficial design in nature, one also finds evidence of chaos and extinction that require revelation to explain. Natural theology could be seen as equating the laws of nature with God's reason and slipping into deism. But an evangelical theology of nature argued that the designs in nature did not provide answers to all religious questions. There was always something hidden, concealed, or mysterious in the functioning of the natural world, and this could only be understood by combining revelation with natural reason (Topham 1999, 144). This combination of natural reason with revelation seems to have led Wilson to the idea of unity in variety which is expressed throughout all of his theology of nature. As we have seen, in his paper on "Poetry and Chemistry," Wilson quotes Bacon as saying: "it has pleased God purposely to conceal his designs from us, that our faculties may be exercised by penetration through the transparent veil" (G. Wilson 1846, 14). Wilson's theology of nature seems to have been aimed at penetrating through the veil in order to discover the unity in variety and to do this would require revelation along with natural reason.

References

Bacon, Francis. 1884. *Bacon's Essays and Wisdom of the Ancients*. Boston, MA: Little, Brown, and Company.

Bacon, Francis. 1937. *Essays, Advancement of Learning, New Atlantis, and Other Pieces*, edited by Richard Foster Jones. New York: Odyssey Press.

Balfour, John H. 1860. *Biographical Sketch of the Late George Wilson, M.D.* Edinburgh: Murray and Gibb.

Barbour, Ian. 1990. *Religion in an Age of Science: The Gifford Lecture*, vol. 1. San Francisco: HarperSanfrancisco.

Bebbington, David. 1999. "Science and Evangelical Theology in Britain from Wesley to Orr." In *Evangelicals and Science in Historical Perspective*, edited by David N. Livingstone, Darryl G. Hart, and Mark A. Noll, 120–141. New York: Oxford University Press.

Brooke, John Hedley. 1974. "Natural Theology from Boyle to Paley." In *New Interactions between Theology and Natural Science*, edited by John Hedley Brooke, Reijer Hooykaas, and Clive Lawless, 8–9. Milton Keynes: Open University Press.

Brooke, John Hedley. 1991. *Science and Religion: Some Historical Perspectives*. Cambridge: Cambridge University Press.

Brooke, John Hedley. 1999. "The History of Science and Religion: Some Evangelical Dimensions." In *Evangelicals and Science in Historical Perspective*, edited by David N. Livingstone, Darryl .G. Hart, and Mark A. Noll, 17–42. New York: Oxford University Press.

Cannon, Susan Faye. 1978. *Science in Culture: The Early Victorian Period*. New York: Science History Publications.

Chalmers, Thomas. 1835. *On the Power, Wisdom, and Goodness of God as Manifested in the Adaption of External Nature to the Moral and Intellectual Constitution of Man*, 2 vols. London: William Pickering.

Crowe, Michael J. 1986. *The Extraterrestrial Life Debate 1750–1900: The Idea of the Plurality of Worlds from Kant to Lowell*. Cambridge: Cambridge University Press.

Desmond, Adrian. 1992. *The Politics of Evolution: Morphology, Medicine, and Reform in Radical London*. Chicago: University of Chicago Press.

Dettelbach, Michael. 1996. "Humboldtian Science." In *Cultures of Natural History*, edited by Nicholas Jardine, James A. Secord, and Emma C. Spary, 287–304. Cambridge: Cambridge University Press.

Draper, John William. 1875. *History of the Conflict between Religion and Science*. New York: D. Appleton and Company.

Gladstone, John Hall. 1860. "Estimate." In *Memoir of George Wilson*, edited by Jessie Aitken Wilson, 509–522. Edinburgh: Edmonston and Douglas.

Herschel, John F.W. 1849. *Outlines of Astronomy*, 2nd ed. London: Longman, Brown, Green and Longmans.

Herschel, John F.W. 1851. *Preliminary Discourse on the Study of Natural Philosophy*, new ed. London: Longman, Brown, Green & Longmans.

Humboldt, Alexander von. 1858. *Cosmos: A Sketch of the Physical Description of the Universe*, vol. 1, translated by E.C. Otte. New York: Harper & Brothers.

Kuhn, Thomas. 1959. "Energy Conservation as an Example of Simultaneous Discovery." In *Critical Problems in the History of Science*, edited by M. Claggett, 321–356. Madison: University of Wisconsin Press.

Livingstone, David, Darryl G. Hart, and Mark A. Noll. 1999. "Introduction: Placing Evangelical Encounters with Science." In *Evangelicals and Science in Historical Perspective*, edited by David Livingstone, Darryl G. Hart, and Mark Noll, 3–17. New York: Oxford University Press.

Rehbock, Phillip F. 1983. *The Philosophical Naturalists: Themes in Early Nineteenth-Century British Biology*. Madison: University of Wisconsin Press.

Richards, Robert J. 2002. *The Romantic Conception of Life: Science and Philosophy in the Age of Goethe*. Chicago: University of Chicago Press.

Secord, James. 2000. *Victorian Sensation: The Extraordinary Publication, Reception, and Secret Authorship of Vestiges of the Natural History of Creation*. Chicago: University of Chicago Press.

Smith, Crosbie. 1998. *The Science of Energy: A Cultural History of Energy Physics* in Victorian Britain. Chicago: University of Chicago Press.

Somerville, Mary. 1840. *On the Connexion of the Physical Sciences*, 5th ed. London: John Murray.

Stanley, Matthew. 2015. *Huxley's Church and Maxwell's Demon: From Theistic Science to Naturalistic Science*. Chicago: University of Chicago Press.

Topham, Jonathan R. 1999. "Science, Natural Theology, and Evangelicalism in Early Nineteenth Century Scotland: Thomas Chalmers and the Evidence Controversy." In *Evangelicals and Science in Historical Perspective*, edited by David N. Livingstone, Darryl G. Hart, and Mark A. Noll, 142–176. New York: Oxford University Press.

Turner, Frank M. 1974. *Between Science and Religion: The Reaction to Scientific Naturalism in Late Victorian England*. New Haven, CT: Yale University Press.

Turner, Frank M. 1993. *Contesting Cultural Authority: Essays in Victorian Intellectual Life*. Cambridge: Cambridge University Press.

Whewell, William. 1837. *History of the Inductive Sciences, from the Earliest to the Present Times*, 3 vols. London: John W. Parker.

Whewell, William. 1840. *Philosophy of the Inductive Sciences*, 2 vols. London: John W. Parker.

Whewell, William. 1845. *Indications of the Creator: Extracts, Bearing upon Theology, from the History and Philosophy of the Inductive Sciences*. London: John W. Parker.

Whewell, William. 1853. *Of the Plurality of Worlds: An Essay*. London: John W. Parker.

Whewell, William. 1855. "Preface to the Third Edition." In *Of the Plurality of Worlds: An Essay*, 3rd ed., edited by William Whewell. London: John W. Parker.

White, Andrew Dickson. 1896. *The Warfare of Science with Theology in Christendom*. New York: D. Appleton and Company.

White, Lynn, Jr. 1967. "The Historical Roots of Our Ecological Crises." *Science* 155: 1203–1207.

Williams, L. Pearce. 1971. *Michael Faraday: A Biography*. New York: Simon and Schuster.

Wilson, George. 1846. "On the Alleged Antagonism between Poetry and Chemistry." *The Torch: Journal of Literature, Science, and the Arts* 1: 13–16.

Wilson, George. 1850a. *Chemistry: An Elementary Text-Book*. Edinburgh: W. & R. Chambers.

Wilson, George. 1850b. *Introductory Address Delivered at the Opening of the Medical School, Surgeon's Hall*. Edinburgh: Sutherland and Knox.

Wilson, George. 1852a. *The Grievance of the University Tests, as Applied to Professors of Physical Science in the Colleges of Scotland: A Letter Addressed to the Right Honourable Spencer H. Walpole, Secretary of State for the Home Department*. Edinburgh: Sutherland and Knox.

Wilson, George. 1852b. *Life of Dr. John Reid*. Edinburgh: Sutherland and Knox.

Wilson, George. 1853. *Five Gateways of Knowledge*. London: Macmillan.

Wilson, George. 1856. *On the Character of God, as Inferred from the Study of Human Anatomy: Addresses to Medical Students Delivered at the Instance of the Edinburgh Medical Missionary Society*. Edinburgh: Adam and Charles Black.

Wilson, George. 1859. *Electricity and the Electric Telegraph, together with The Chemistry of the Stars*, 3rd ed. London: Longman, Brown, Green, Longmans, & Roberts.

Wilson, George and Archibald Geikie. 1861. *Memoir of Edward Forbes, F.R.S.* London: Macmillan.

Wilson, George. 1862a. "Chemical Final Causes: As Illustrated by the Presence of Phosphorous, Nitrogen, and Iron in the Higher Sentient Organisms." In *Religio Chemici: Essays*, edited by George Wilson 104–164. London: Macmillan & Co.

Wilson, George, 1862b. "Chemistry and Natural Theology." In *Religio Chemici: Essays*, edited by George Wilson, 1–50. London: Macmillan & Co.

Wilson, George. 1862c. "The Chemistry of the Stars: An Argument Touching the Stars and Their Inhabitants." In *Religio Chemici: Essays*, edited by George Wilson, 51–103. London: Macmillan & Co.

Wilson, George. 1862d. *Counsels of an Invalid: Letters on Religious Subjects*. London: Macmillan & Co.

Wilson, George. 1862e. "The Life of Wollaston." In *Religio Chemici: Essays*, edited by George Wilson, 253–303. London: Macmillan & Co.

Wilson, Jess Aitken. 1860. *Memoir of George Wilson*. Edinburgh: Edmonston and Douglas.

Wilson, Jessie Aitken. 1866. *Memoir of George Wilson*, new ed. London: Macmillan and Co.

4   Wilson's Methodology of Science

A theology of nature and the idea of unity in variety provided George Wilson with an ideology of science but there was still the question of what methodology could be used to actually discover that unity. Although almost all of his non-technical writings make at least some passing reference to God, much of his writings on secular subjects focus on issues associated with the methodology of science. Wilson's methodology of science seems to be influenced by two interconnected sets of ideas – Scottish Common Sense philosophy and Baconian philosophy. As we have seen as a student Wilson was reading John Abercrombie, a leading Common Sense philosopher, who he said had "the highest reputation among medical men," and he referred to F.D. Maurice, another leading Common Sense philosopher who he said was the "most accomplished writer of the age" (G. Wilson 1862c, 251; J. Wilson 1860, 367). In addition, John Cairns, who Wilson called his spiritual father, studied at Edinburgh with William Hamilton whose goal was to reconcile Common Sense philosophy with Kantian transcendentalism. As we will see, much of Wilson's writings place emphasis on the role of the senses, which was a key element of Common Sense philosophy.

Baconian Philosophy

Probably, the most important influence on Wilson's methodology of science was Francis Bacon. He mentions Bacon often in his writings, and in his paper "On the Alleged Antagonism between Poetry and Chemistry," which was discussed in the last chapter, Wilson calls Bacon "the great father of modern science" (G. Wilson 1846, 14). John Cairns, in his obituary of Wilson, says: "His heart was strung throughout in sympathy with the touching prayers of the *Novum Organon*, (*sic*) that science become a healing art..." (Cairns 1860, 202). Bacon's ideas were especially popular and influential in early Victorian Great Britain (Yao 1985). A number of the Scottish Common Sense philosophers drew on Bacon for many of their ideas because of his emphasis on the role of the senses. Thomas Reid, one of the founders of Common Sense philosophy, was drawn to Bacon because of his emphasis on the senses, which was a key element of Common Sense philosophy, and said that he was "very apt

DOI: 10.4324/9781003212218-4

to measure a man's understanding by the opinion he entertains of that author [Bacon]," and Dugald Stewart, one of the leading Common Sense philosophers in the nineteenth century, called Bacon the "father of Experimental Philosophy" (Yao 1985, 260). In England, Samuel Taylor Coleridge, who was influential on Forbes's Brotherhood, referred to Bacon as the "British Plato" (Yao 1985, 257). John Herschel's *Preliminary Discourse on the Study of Natural Philosophy* (1831) said that before "the publication of the Novum Organum of Bacon, natural philosophy, in any legitimate and extensive sense of the word, could hardly be said to exist" (Yao 1985, 267–268). Bacon's philosophy had also become a central focus of William Whewell's *History of the Inductive Sciences, from the Earliest to the Present Times* (1837) and *Philosophy of the Inductive Sciences, founded upon Their History* (1840), which were seen by him as attempts to both follow and reform Baconian philosophy (Olson 1975, 34, 94; Whewell 1840; Whewell 1837). In his *Philosophy* Whewell claimed that Bacon was "not only one of the Founders, but the supreme Legislator of the modern Republic of Science" (Yao 1985, 273).

One problem in dealing with Bacon's influence on nineteenth-century British science was that his reputation was undergoing change throughout the period (Yao 1985, 251). During the seventeenth and eighteenth centuries, Bacon's philosophy was as extending beyond a reform of science to also a reform of society. Bacon's Salomon's House in the *New Atlantis* became the model for the Royal Society of London, and Bacon was seen as a supporter of a new model of cooperative science that through its experimental nature could be more democratic, and his emphasis on the utility of science would connect science to the mercantile and industrial classes rather than the university-trained elite (Yao 1985, 254–255; Webster 1974). This made Bacon's works influential in eighteenth-century France among the Encyclopedists and in democratic nineteenth-century Scotland, but this also was of some concern in England which saw the association with French political ideas as a danger. Bacon's emphasis on the senses could also be seen as support of materialism and atheism (Yao 1985, 258). Bacon's idea that the role of science should be "the relief of man's estate" met some resistance in England where people like Thomas Macaulay warned of Baconianism satisfying "vulgar wants" (Yao 1985, 258). This idea of "vulgar Baconianism" also seemed a threat to the British scientists who were trying to professionalize and raise the status of science through the establishment of the British Association for the Advancement of Science in 1831, which took its name from Bacon's *Advancement of Learning*. As a result of this goal to professionalize science by emphasizing the pursuit of pure knowledge, the focus shifted to Bacon's methodology, such as that presented in the *Advancement of Learning* or the *Novum Organum*.

There has often been a great deal of misunderstanding about Bacon's methodology (Zagorin 1998; Martin 1992; Briggs 1989; Rossi 1978). It is often assumed that his methodology was nothing more than pure or naïve empiricism, where one simply collects facts and then through a process of induction one is somehow able to generalize from those facts and discover general laws.

This has come to know as the "Baconian method" and is also assumed to reject all hypotheses and completely rely on information from the senses, but his methodology is much more complex than often presented. Bacon's main goal was to bring about a reform in the way we gain knowledge in order to improve the human condition. He was concerned that the traditional Aristotelian method as taught in the universities had not kept pace with the changes brought about by new inventions, such as the printing press, gunpowder, and the compass (Zagorin 1998, 35). In order to accomplish his reform of methodology, Bacon focused on rejecting the authority of Aristotle and his contemplative approach to philosophy and replacing it with a more active approach that would connect the knowledge of something with knowing how to make that thing (Zagorin 1998, 38). Bacon believed that such a new approach to philosophy would lead to improvements in human life, but in order to accomplish this new approach, he would first have to establish a new methodology toward gaining knowledge.

Bacon presented his ideas in a number of works, including *The Advancement of Learning*, the *Novum Organum* (or new tool for learning), and the *New Atlantis*. Especially, in the *Novum Organum*, Bacon put forward the idea that an understanding of natural philosophy had to begin with empirical data, then proceed through a process of induction to an understanding of the material and efficient causes of phenomena, then to a metaphysical understanding which culminated in the discovery of laws of nature (Zagorin 1998, 64). He argued that the process should begin with what he called "natural histories" of all phenomena which would provide the data to which the inductive process would be applied (Zagorin 1998, 103). Bacon has often been portrayed as advocating a simple or naïve collection of facts, but he viewed his "natural histories," as depending much more upon an experimental approach to nature rather than simple observations. For Bacon, the important data about nature "cannot appear so fully in the liberty of nature as in the trials and vexations of art" (Zagorin 1998, 62). That is, like the lawyer that he was Bacon believed that nature did not give up her secrets without some form of interrogation.

In order to obtain axioms, which for Bacon included such things as causal explanations, theories, or laws of nature, a new form of induction would have to be used (Zagorin 1998, 87–88). Again, there has been a great deal of misunderstanding concerning Bacon's idea of induction. Many twentieth-century philosophers criticized Bacon for misunderstanding how science actually operates, because they believed that he neglected the role of hypothesis in the development of scientific theories and relied instead on the idea that laws could be discovered through simple generalizations based on observable facts. But more recent scholarship on Bacon has shown that his method was more complex than simple induction. Bacon advocated drawing up tables of presence (lists of examples when a phenomenon was present), tables of absence (lists of examples when a phenomenon was not present), and tables of comparison (lists of when a phenomenon increased or decreased in the previous tables). Then by using the tables to exclude explanations for phenomena,

Bacon would be left with the true explanation of a phenomenon. Although such an approach could be labeled inductive, it was never the undirected process that modern critics have claimed. According to one scholar, the so-called axioms that Bacon was searching for in his process "included the concept of a theory of hypothesis to be tested and corroborated by its prediction and discovery of new facts and observations" (Zagorin 1998, 88). Bacon himself always argued in favor of a combination of empiricism and rationalism. In the *Novum Organum*, he criticized both pure empiricists and pure rationalists calling the rationalists spiders who spun theories out of their own bodies and called empiricists ants who simply piled up facts. Instead of the ant or the spider, Bacon's goal was to be a bee that collected nectar but transformed it by its own power into the useful product of honey. All this has led at least one scholar to classify Bacon's method as closer to a hypothetic-inductive method (Zagorin 1998, 101–102).

As we have seen, Susan Faye Cannon has argued that during the early nineteenth century, scientists were no longer following a Baconian method, but more recent scholarship has revised her view (Yao 1985, 269). Although Bacon began to fall out of favor during the second half of the century, during the first half of the century he was still held in high esteem among many British scientists and philosophers, as we have noted above, but there were some questions raised as to the originality of Bacon's inductive method (Yao 1985, 260–263). More direct attacks came from historian/politician Thomas Macaulay and physicist David Brewster (Yao 1985, 266–272). Much of Macaulay's criticism of Bacon focused on his character and his time as a politician. Brewster did criticize Bacon's inductive method saying that it ignored how actual scientists made discoveries (Yao 1985, 278–280). While not directly criticizing Bacon's inductive method, a number of philosophers and scientists began to make revisions to that method. In particular, Thomas Brown and Dugald Stewart, two of the leading Common Sense philosophers, began to move away from Thomas Reid's warnings against the use of hypotheses and argue that hypotheses could be useful as a first tentative step toward the development of discovery and could help to guide experiments but still warned against accepting hypotheses as reality unless confirmed by experiments (Yao 1985, 264–265). John Herschel in his *Preliminary Discourse on the Study of Natural Philosophy*, a work that Wilson recommended to a young friend, sought to emphasize the importance of the Baconian method in the verification of discoveries rather than in the discoveries themselves (Yao 1985, 269; G. Wilson 1862a, 6). Whewell, another practicing scientist like Herschel, had also praised Bacon, as we have seen, but he was also critical of some aspects of Bacon's method. He argued that Bacon was not simply a pure empiricist but "had the merit of showing that Facts and Ideas must be combined" (Yao 1985, 273). But Whewell went on to criticize what he saw as a mechanical process of discovery that ignored the "role of the inventive genius, which all discovery requires" (Yao 1985, 275). Whewell's theory of induction, which he saw as a revision of Bacon's, gave a much greater role

to the human mind in giving some order to the facts gained through observation. Finally, the reaction to Bacon often differed from discipline to discipline. By the middle of the nineteenth century, many physicists, such as
David Brewster, were abandoning Bacon. This may be because physics was
becoming more theoretical and mathematical, and Bacon had not addressed
those issues. On the other hand, the more naturalistic sciences, like geology
and paleontology, still found Bacon useful since his inductive method was
one that they were using, and it helped give to legitimate and professionalize
sciences like geology that had been dominated by religious ideology and theories such as catastrophism during much of the eighteenth century (Yao 1985,
287). In particular, the newly formed Geological Society of London, with
which Wilson's friend Edward Forbes had been associated, embraced the
Baconian program as a way to give authority to their work. Bacon's approach
also allowed the new developing field of geology to draw on a wide range of
observers to collect geological data.

Biographies

Bacon's ideas and his methodology show up explicitly in Wilson's scientific
biographies which became some of Wilson's most influential non-technical
writings. John Balfour said of Wilson's biographical memoirs that "he shone
with marked luster" and goes on to say of his biography of Henry Cavendish:
"It is an admirable biography, 'full of life, or picturesque touches, and of realization of the man and of his times,'" and says of Wilson's biography of the
Scottish physiologist John Reid, that it

> is a vivid and memorable presentation to the world of the true linea-
> ments, manner of life, and inmost thought, and heroic sufferings, as well
> as the noble scientific achievements of the strong, truthful, courageous,
> and altogether admirable man and true discoverer, — a genuine follower
> of John Hunter.
>
> (Balfour 1860, 24–25)

Wilson notes that during "the enforced leisure of a long illness, I commenced
in 1842, to collect materials for a projected work on the lives of the Chemists
of Great Britain ..." (G. Wilson 1851, v). This was the year that Wilson was
suffering from the ankle infection that would lead to the partial amputation
of his foot the next year. Before he could complete the project, Wilson was
contacted by members of the Cavendish Society. The Cavendish Society had
been formed in 1846 with the object of "translating and publishing works
on chemistry and its applications" (Provincial Medical and Surgical Society
1846, 446–447). Its President was Thomas Graham, in whose laboratory
Wilson had worked in London, and members of the council were Michael
Faraday and Lyon Playfair, who had also worked with Wilson in Graham's
laboratory. One of the first books the Society commissioned was a biography

of Henry Cavendish, which Wilson agreed to write (G. Wilson 1851, v). Although Wilson put aside his larger project of the lives of British chemists, he did publish three shorter sketches of John Dalton, Henry Wollaston, and Robert Boyle in the *British Quarterly Review* in 1845, 1846, and 1849 (G. Wilson 1862b, 1862c, 1862d).

Wilson does not specifically say why he became interested in scientific biographies, but in the mid-1840s, there seemed to be an increased interest the scientific biographies (Jungnickel and McCormmach, 1999, 10). In 1845, Henry Lord Brougham complained that biographies had ignored many British scientists and he sought to remedy this by publishing *Lives of Men of Letters and Science* which included brief biographies of Joseph Black, James Watt, Joseph Priestly, Henry Cavendish, and Humphrey Davy (Brougham 1845, xi). There is also some evidence that Wilson was following Bacon's call for the need to create "natural histories" in order to have data to which the inductive process could be applied. In each one of his biographies, Wilson mentions either the importance of Bacon, the importance of induction or both. In his biographical sketch of Robert Boyle, Wilson specifically states that he sees Boyle as the intellectual heir of Bacon and that is how Wilson will treat him (G. Wilson 1862b, 179). In that same sketch, Wilson seems to be following Bacon's argument that nature did not give up her secrets without some methodology to extract them when he quotes the bible saying "It is the glory of God to conceal a thing, but the honour of kings to search out a matter" (G. Wilson 1862b, 172). As we have seen in his article "Poetry and Chemistry," Wilson had earlier referred to Bacon's use of the same passage. Wilson may have also been drawn to the history of science by William Whewell's publication of *History of the Inductive Sciences, from the Earliest to the Present Times* (1837). One of Whewell's goals was to bring about a reform of Bacon's philosophy which would become the focus of his *Philosophy of the Inductive Sciences, founded upon Their History* (1840). But, following Bacon, Whewell believed that philosophy of science had to begin with its history and that Bacon had not shown exactly how induction had been used by practicing scientists in the past. Wilson was often ambivalent when it came to Whewell. As we have seen, he seemed to welcome Whewell's support of his paper on dealing with the plurality of worlds, and in an 1847 letter to a young friend, he recommends reading Whewell's *Bridgewater Treatise* as well as his works on the history and philosophy of the inductive sciences (G. Wilson 1862a, 6). But Wilson was critical of Wilson's *History of the Inductive Sciences* in public and even more so in private. In his "Life and Discoveries of Dalton," Wilson claims that Whewell did not understand how Dalton's atomism was founded on experimental evidence, and in a letter to his friend and publisher, Daniel Macmillan, Wilson says of Whewell: "His presumptuous history of the Inductive Sciences is wofully [*sic*] shamefully inaccurate" (G. Wilson 1862b, 322; Miller 2004, 209–210). Wilson was probably sympathetic to the idea of a history of the inductive sciences, but the defects he saw in Whewell's work may have stimulated him to write his own histories.

John Dalton

George Wilson's first work on the history of chemistry was his "Life and Discoveries of Dalton," published in the *British Quarterly Review* in 1845. Wilson notes that no previous study of Dalton's life or scientific work had been done. In 1854, William Charles Henry wrote *Memoirs of the Life and Researches of John Dalton* and in the preface called Wilson's article "elaborate and well-conceived," and "beyond comparison the ablest and justest appreciation that has appeared of Dalton's philosophical character and discoveries" (G. Wilson 1862b, 304; Henry 1854, x). Wilson's study of Dalton begins and ends with some relatively brief biographical materials and what might be expected, a discussion of Dalton's color blindness, but the main part of the work is a somewhat detailed study of the steps and methodology that led to Dalton's "Atomic Theory." Wilson may have been especially interested in this aspect of Dalton's scientific work since, as we have seen, Wilson himself seemed to be drawn to atomism in chemistry. As noted, he was both friends and a great supporter of Samuel Morrison Brown and what Brown called transcendental chemistry that saw all matter composed of identical atoms and that all of the known chemicals were simply different configurations of these identical atoms (Rehbock 1983, 87–90). We have also seen that Wilson spent significant time trying to reproduce Brown's controversial claim that he had transmuted carbon into silicon. Wilson may have also been drawn to Dalton because he interpreted his work as reflecting the unity that Wilson saw as fundamental in nature. Wilson says:

> Great unity, and the impress of intellectual consistency, are stamped on all Dalton's labours. With few exceptions, they bear closely and directly upon each other, and on the atomic hypothesis of combining proportions, to which they ultimately led, and round which they naturally group themselves.
>
> (G. Wilson 1862b, 309)

Wilson then goes on to indicate that his aim in his study of Dalton is to focus on the method by which Dalton led to the unity of the atomic theory. He says: "The method which we shall follow, will serve, accordingly, both to bring out the nature and value of his discoveries in science, and to indicate the train of speculation and inquiry by which he was conducted to them" (G. Wilson 1862b, 309).

Although Wilson does not specifically mention Bacon, his presentation of Dalton's discovery of the atomic theory reflects and follows a very Baconian approach to science. He begins by noting that Dalton's early interest in meteorology was the source of all of his great discoveries and that this interest led Dalton in 1788 to begin making detailed daily observations of meteorological data such as temperature, barometric pressure, and wind speed. He would continue to make daily observations up until the night before he died, making more than 200,000 over a period of about fifty years (G. Wilson 1862b, 307, 354).

Wilson notes that beginning with meteorology was important since it involved aspects of almost all of the other physical and chemical science. In his early days living in the small town of Kendal, Dalton had access to only a small library, so he began his study of the problems that his meteorological observations had raised by conducting experiments. This was certainly a very Baconian approach to scientific research. Once he moved to Manchester, Dalton began to focus his research on questions relating to heat and the physical constitution of gases, both of which were crucial to understanding meteorological phenomena, especially how different chemical elements mix together in the atmosphere. In conducting research on gases, Dalton was able to induce four laws of the proportional combination of different chemical elements. Some of these laws had been discovered by previous chemists but some were the result of Dalton's research. Wilson notes that "[t]hese laws, it is important to observe, contain in them nothing hypothetical. They sum up the results of the universal experience of chemists (so far as experience can be called universal), of which they are expressions" (G. Wilson 1862b, 323). Dalton's next step was to introduce two hypotheses that would form the basis of his atomic theory. First, he argued that gases (and all other material) were made up of atoms. This was not a radically new idea. The idea of atoms can be traced back to the Greeks and by Dalton's time most scientists, including Newton, Robert Boyle, and many others assumed that gases were composed of small atom–like particles. Dalton's second hypothesis was radical and unique to Dalton. He assumed that the atoms of different elements had different weights (G. Wilson 1862b, 340–341). With this hypothesis, he was able to explain the four laws of proportional combination and provided a new model for understanding chemistry.

It might be argued that Dalton's use of an atomic hypothesis was a violation of the Baconian method, but it was only a violation of the commonly misunderstood Baconian method. Bacon himself accepted atomism, and it was an important element of his inductive method. He notes that the goal of induction was not to explain nature by abstractions but "to dissect her into parts; as did the school of Democritus, which went further into nature than the rest" (Bacon 1937, xxiii). In fact, Bacon was trying to prove the idea that heat was caused by the motion of particles, when he conducted the experiment that led him to catch a cold and die. In any case, Wilson argues that Dalton's use of an atomic theory was part of an inductive process. Wilson says:

> To Dalton himself the evidence in support of the existence of ultimate indivisible particles appears to have seemed so conclusive, that he considered the doctrine of atoms in the light of an induction from the data furnished by observation and experiment; and this without reference to any other physical question.
>
> (G. Wilson 1862b, 342)

Wilson also presents Dalton's development of an atomic theory in light of other ideas that Wilson sees as important in science. He argues that Dalton's researches "had doubtless strengthened that faith in the uniformity of nature's laws which we all inherit as an essential part of our mental constitution" (G. Wilson 1862b, 341). Here, we see Wilson making a connection between one of his favorite ideas – unity – with the Common Sense philosophy idea that the mind has intuitive or instinctual aspects that allow it to interpret information from the senses. Near the end of his work, Wilson also connects Dalton's work to a vision of science held by both Wilson and Bacon – the idea that science is useful and utilitarian. He says:

> The application of the laws of combining proportion to the practical arts enabled the manufacturer of glass, of soap, of pigments, of medical substances, or dyes, of oil, of vitriol, and of many other bodies of great commercial value, to secure their production without waste, or loss, or any unnecessary expenditure.
>
> (G. Wilson 1862b, 352)

William Hyde Wollaston

Wilson's second short biography was of the British chemist William Hyde Wollaston who gained his major fame as the discoverer of palladium and rhodium. It was published in the *British Quarterly Review* in 1846 (G. Wilson 1862d, 253–303). Wollaston had a wide range of interests, including astronomy crystallography, optics in addition to organic and inorganic chemistry, and even "Fairy Rings." Along with Humphrey Davy and Thomas Young, all of whom died a few months apart, Wollaston was part of the period in the early nineteenth century that has been called the "Golden Age of Science" (Wayling 1927, 81). Wilson notes that while much had been written about Humphrey Davy, there was not an even brief biography of Wollaston. In fact, it would not be until 2015 that a modern biography would be published (Usselman 2015). The fact that Wilson did not have access to Wollaston's personal papers and laboratory notebooks limits Wilson's study to Wollaston's published works and some brief biographical details. Like Wilson, Wollaston began studying medicine and practiced it briefly until he inherited enough money to allow him to pursue his scientific interests that were wide-ranging.

Although Wilson does not specifically mention Bacon, his biography reflects certain Baconian themes and methods. Wilson sees the center of Wollaston's scientific life as his laboratory which he would not allow even his closest friends to enter. Wilson focuses much of his study on two of Wollaston's publications – one describing a technique to make platinum malleable and the other a description of his reflecting goniometer (G. Wilson 1862d, 254). Both of these played an important role in expanding an experimental tradition in chemistry and crystallography. The ability to make platinum malleable gave analytic chemists a whole new set of tools (G. Wilson 1862d, 264–268). Before Wollaston's discovery, chemists who wanted to analyze acids, alkalis, and other solvents were limited to porcelain or glass containers, but these were not useful if the substances needed to be heated to a high temperature. Platinum, which had been discovered about 1750 was known to have a very high melting point and would resist chemical action from almost all other chemicals, but it existed in the form of small grains that resisted being shaped. Through what appears to be a purely experimental study, Wollaston developed a process of chemical purification, then powerful compression, and then heating and hammering that resulted in a malleable form of platinum that could be shaped into crucibles for use in chemical analysis. This process made Wollaston a very wealthy man. Wilson estimates that his process brought him more than 30,000 pounds (G. Wilson 1862d, 257).

Wollaston's other major contribution to experimental science was his invention of a reflecting goniometer (G. Wilson 1862d, 269–276). The goniometer, which had existed before Wollaston, was a device to measure angles, particularly the angles of crystals but most were not very accurate and could only be applied to large crystals. Wollaston's improvement was to use reflected light from the faces of the crystals to measure angles rather than physically measuring the crystals themselves. This was much more accurate and could be applied to the smallest crystals. For Wilson, this was an important invention because he saw crystals as a unifying idea since they arose in chemistry, mineralogy, optics, and medicine and were useful in the study of heat and magnetism as well as a topic of study for mathematicians. In fact, it is in discussing the study of crystals that Wilson makes one of his uses of the phrase "Unity in Variety" (G. Wilson 1862d, 274). As an example, Wilson notes that the extended observations using Wollaston's goniometer resulted in the discovery that while many different crystals can have the same shape, such as an octahedron, in the same substance the angles of the faces will always be the same, while in a different substance that same shape will have a different angle. It was further discovered that when substances are very similar, their crystalline form, or shape, is very similar. This led to the discovery of isomorphism in which two different closely related chemicals each form a compound with a third chemical, resulting in two different compounds that have the same crystalline form. Wilson goes on to note that these discoveries came purely from experiments and observations (G. Wilson 1862d, 274–275). While not specifically mentioning Bacon, Wilson's description of finding unity in nature through observation and experimentation seems to be an example of the Baconian method. He also goes on to argue that crystalline form has become one of the most important ways to classify chemical substances, and "to arrange them into groups and natural families," which also reflects Bacon's method (G. Wilson 1862d, 275). This connection to Bacon is reinforced when Wilson compares Wollaston's approach to science to that of Davy. He notes that Wollaston excelled Davy when it came to "minute accuracy of observation," and that he differed from Davy in that "Wollaston's idea of truth was not so much something proved true, but something which could not be proved to be not true" (G. Wilson 1862d, 291). This last observation is interesting since in the *Novum Organum* Bacon says: "In the process of Exclusion are laid the foundations of true Induction ..." (Bacon 1937, 341). That is, for Wollaston, as for Bacon, it was important to know what was not true in order to discover what was true. Wilson also seems to connect Wollaston with Bacon when he says:

> Wollaston's genius was like the light, whose laws he so much loved to study. It was not, however, the blazing light of day that it resembles, but the still moonlight, as ready with clear but cold radiance to shin in on a solitary obscure chamber, as able to illuminate with its unburning beams, every dark and stately hall of the closed fortresses where Nature keeps her secrets.
>
> (G. Wilson 1862d, 292)

We have already noted in the last chapter that Wilson on more than one occasion made reference to Bacon's *lumen siccum*, or cold or dry light or reason, and also referred to Bacon's saying that "it has pleased God purposely to conceal his designs." Finally, Wilson comments on the "acuteness of Wollaston's senses as the source of his greatness as an inventor and discoverer" (G. Wilson 1862d, 301). Here, it is possible that Wilson is again arguing for the role of Common Sense philosophy in the development of a methodology of science. In any case, it could be possible to interpret Wilson's study of Wollaston as an example of how the application of something like a Baconian methodology can lead to the discovery of the ideology of the unity of nature.

Robert Boyle

The longest of Wilson's "Sketches" was the "Sketch of the Life and Works of the Hon. Robert Boyle" which was published in 1849 in the *British Quarterly Review* (G. Wilson 1862c, 165–252). This sketch is also the most explicitly Baconian. Boyle himself claimed that he avoided grand systematic schemes, including those of Gassendi, Descartes, and even Bacon, in order to not corrupt his experiments, but a number of recent historians of science have questioned the veracity of this claim (Shapin and Schaffer 1985, 68–69, 68n). In any case, Wilson sees Boyle clearly within the Baconian program and begins his essay with a discussion of the early history of the Royal Society of London. While crediting King Charles II with formally establishing the Society, he says he simply provided a name but was not the founder, saying:

> If any person can claim that honour, it is Lord Bacon, who, by the specific suggestions in his *New Atlantis*, but also we believe still more, by the whole tenor of his *Novum Organum*, and other works on science, showed his countrymen how much can be done for its furtherance, by the co-operation of many labourers.
>
> (G. Wilson 1862c, 170)

He goes on to note that following Bacon's argument that science needed to be useful, the Royal Society's discoveries and inventions have played a direct role in the commercial prosperity of the kingdom, especially in the area of navigation. Wilson argues that societies, such as the Royal Society, arose when scientists began to think that "the robes which God had flung over the naked-ness of the material world, might be worth looking at, and might prove a more glorious apparel than the ideal garments which man's imagination had fashioned for the universe" (G. Wilson 1862c, 172). This is a very Baconian argument. First, it repeats Wilson's often reference to Bacon's idea that God has concealed his designs and it is the purpose of humans to penetrate through the veil. It also argues that a direct study of nature is better than humanly con-structed models or theories. Wilson is explicit that he is placing Boyle within this historical context and sees Boyle as the heir to Bacon saying:

> It was long ago observed that Boyle was born in the year in which Bacon died, and it soon appeared that a corner, at least, of the deceased prophet's mantle had fallen upon him. He was the earliest pupil who applied, in practice, the lessons of the *Novum Organum* … As patriarch, therefore, of English experimental science, he takes precedence even of Newton. It is in this capacity that we propose chiefly to treat of Boyle.
>
> (G. Wilson 1862c, 179)

Wilson provides some details of Boyle's early life, including being stranded on the Continent for two years with his older brother waiting for funds to return to Great Britain, while his father had to spend his money to pay troops to put down a general rebellion in Ireland. On his return, Boyle turned to a study of science and eventually became famous for his work in a number of sciences, including chemistry, pneumatics, heat, acoustics, and even med-icine. But the focus of Wilson's "Sketch" is Boyle's experimental researches using the air-pump. Wilson provides a detailed and extensive history of the development of the air-pump in England and also in 1849 he published a separate "On the Early History of the Air-Pump in England" in *the Edinburgh New Philosophical Journal* (G. Wilson 1849). Both of these works have received praise from modern historians of science Steven Shapin and Simon Schaffer in their book *Leviathan and the Air-Pump* saying: "The best overall accounts [of Boyle's air-pumps] remain the nineteenth-century essays of Wilson …" (Shapin and Schaffer 1985, 26n). Wilson clears up a number of mispercep-tions and confusions about Boyle's air-pumps (G. Wilson 1862c, 190–218). After hearing of Otto von Guericke's invention of an air-pump in Germany, Boyle became interested in constructing an air-pump in order to use a vac-uum to conduct experimental research, but Boyle himself did not have the skill to construct such a device so he had to rely on Robert Hooke, his assis-tant, to build the first air-pump in England which was completed in 1659. A few years later, in 1667, a second air-pump was constructed with the help of Hooke, and a dramatically redesigned and improved air-pump was also built around 1676 by his new assistant, Denis Papin. So, Boyle's fame would come from using the air-pump as a new scientific tool, not actually inventing or constructing the air-pump.

In the Baconian spirit of collaborative research, Boyle donated his first air-pump to the Royal Society in 1662 so that its members could make use of it to conduct experiments. This has also led to some confusion that Wilson is able to resolve using a fairly detailed historical scholarship. Beginning in the nineteenth century, the Royal Society began to display an air-pump that was claimed to be the original air-pump that Hooke built in 1659, but Shapin and Schaffer credit Wilson for proving that the displayed air-pump could not have been the original of Hooke and Boyle or even the improved air-pump of Papin (Shapin and Schaffer 1985, 227).

While Boyle has to share credit with Hooke and Papin for the construction of the air-pump, Wilson claims that Boyle deserves full credit for the researches he conducted with it. The "inventive" experiments Boyle did using the air-pump included work on the mechanical properties of the atmosphere, the life-sustaining properties of the atmosphere, and even converted his pump into a retort, alembic or still to study the chemical properties of gases by liquifying them (G. Wilson 1862c, 219). He also used the pump to extend the work of other scientists, such as Evangelista Torricelli and von Guericke, on the pressure of the atmosphere which would result in establishing the relationship between the pressure and volume of gas in what came to be known as Boyle's Law (or the Boyle-Mariotte Law). In one of his famous experiments, he was able to settle the debate about whether the sound was carried by the air. In some previous experiments done by others, it appeared that a sound produced by a vibrating body placed inside a vacuum could be heard outside the vacuum, but this turned out to be the result of the vibrating body coming into contact with the glass enclosing the vacuum. Even if the air carried sound, there were questions about how it did so. Von Guericke thought that the air carried sound like it did smoke rather than providing a medium to transfer vibrations. Wilson argues instead that sound moves through air like a wave. As noted in an early chapter, Wilson almost copies exactly Mary Somerville's explanation of sound as like a wind moving through a cornfield where the individual stalks of corn only move a small amount, but the wave of wind moves from one end of the field to the other. Following what might be interpreted as Baconian approach, Wilson argues that Boyle's proof that sound was carried by the air did not depend on his accepting any particular theory of sound saying, "he was one whom no theory would prevent from subjecting to direct trial, what he thought experiment only could decide" (G. Wilson 1862c, 221–222). In Boyle's experiment, he hung a ticking watch in the center of the glass globe attached to his air-pump without any contact with the glass. As the air was removed, the sound of the watch diminished until it became inaudible after all of the air was removed, thus proving sound to be carried by the air. In a similar fashion, Boyle was able to show that combustion required air. He placed a number of combustibles, including candles and even pistols, inside his glass globe and showed that combustion would cease or would not start when the air had been removed. Again, following a purely experimental approach, Boyle did not associate his finding with any specific theory of combustion. Ironically, if he had, he might have helped to undermine the theory of phlogiston as some imponderable source of fire a hundred years earlier than the work of Joseph Priestley and Antoine Lavoisier.

Some of Boyle's most controversial experiments with the air-pump had to do with the relationship between air and life, or between combustion and life. Through numerous experiments, he was able to show that air was required to sustain animal life just as it was required to sustain combustion. The experiments used to confirm this fact were often cruel since they involved depriving a large number of different animals of air until they suffocated. One of the more famous paintings of the eighteenth century was a 1768 work by Joseph Wright of Derby entitled, "An Experiment on a Bird in an Air Pump," which now hangs in the National Gallery in London. In the painting, a young girl covers her face in tears as a white bird in the glass globe of an air-pump is clearly being suffocated. As an avowed animal lover, Wilson seems to be particularly concerned with this aspect of Boyle's work. Although Wilson argues that Boyle may have been thinking more about the beneficial aspects of these experiments in the treatment of diseases like tuberculosis, he seems bothered by the fact that Boyle never expressed concern about the suffering while Hooke, who carried out many of the experiments, expressed great remorse and said that he would "never repeat so cruel a deed" (G. Wilson 1862c, 225).

In the last part of the "Sketch," Wilson concentrates on Boyle's character and his religion. With the exception of a brief mention of Dalton's Quaker faith, Wilson's "Sketch" of Boyle is the only one where he discusses religion in any detail. Like Wilson himself, Boyle underwent a "conversion experience," although at a younger age and not associated with a life-threatening medical problem. At age fourteen, while on his extended stay on the Continent with his brother, Boyle was awakened by a violent thunderstorm in Geneva and became convinced that the apocalypse was near and that he was not ready for it. The fear of this episode led to his conversion the next morning. Although having undergone a "conversion experience," Boyle, while still on the Continent, began to have some doubts about Christianity and fell into despair over what he saw as Satanic temptation and appears to have even contemplated suicide (Shapin and Schaffer 1985, 318; G. Wilson 1862c, 235–239). Wilson reports that Boyle would suffer episodes of melancholia for the rest of his life, but by the time he returned to England, he felt that he was back in God's favor and claimed his aim to be to serve God and to serve man. Wilson labeled Boyle a Christian philosopher and noted that he studied nature

> as the Living One, the impress of whose finger he had found on every material object he had examined, 'whose ways' he better than most

men knew 'were past finding out,' but whose works he had found 'all to praise him'.

(G. Wilson 1862c, 248–249)

This seems to be a reflection of Wilson's own belief that a study of nature can find God's designs but not God's true purpose. Wilson did not talk in any detail about Boyle's religious writings saying they are "altogether unsuited to the test of the present day" (G. Wilson 1862c, 249). But the overall focus of Wilson's "Sketch" of Boyle again seems to be to provide an example of a Baconian approach to science.

Henry Cavendish

Wilson's most significant book on the history of science, and the one that garnered him significant fame, was his *The Life of Henry Cavendish*, published in 1851. Christa Jungnickel and Russell McCormmack, the modern biographers of Cavendish, note that the book is not a traditional biography, and they object to Wilson's portrayal of Cavendish as a "man without a heart," and "passionless" (Jungnickel and McCormmach 1999, 12). They attribute this to the very religious Wilson being unable to find that Cavendish had any religious feelings, but a number of other studies have commented on Cavendish's very odd personality, some, including one by the neurologist Oliver Sacks, going so far as to ask the question if he had Asperger's syndrome (Sacks 2001, 1347). In any case, Wilson's *Life of Henry Cavendish* is certainly not organized as a traditional biography. Of the almost 500 pages, only about 100 actually focus on Cavendish's life. Another 80 pages are abstracts of Cavendish's most important papers in chemistry and natural philosophy, but almost 300 pages focus on a detailed study of what was known as the "Water Controversy" or the "Water Question," which involved a very complex debate over who was the first to recognize that water was not an elemental material but a compound of hydrogen and oxygen – was it Cavendish, or James Watt, or Antoine Lavoisier? (Miller 2004).

Wilson traces Cavendish's family back to the time of William the Conqueror and shows that Cavendish as the grandson of the Duke of Devonshire grew up in a family of great wealth, although he did not seem to benefit from that wealth until his father died. Cavendish attended Cambridge but left a few weeks before graduation. Wilson suggests that he did not want to take the rather stringent religious test that was required for completion of the bachelor's degree (G. Wilson 1851, 181). While Wilson does not find any indication that Cavendish followed any religious practices, he speculates that he might have leaned toward Unitarianism and did not subscribe to the doctrine of the trinity. In any case, Cavendish left the University for London, where he spent much of his life pursuing a program of independent scientific research. While he obtained a number of honors and was closely involved with the Royal Society of London, he never held a regular academic position.

Cavendish conducted research into a wide range of scientific areas, including natural philosophy and chemistry, along with geology, meteorology, astronomy, and mathematics. Wilson clearly sees Cavendish's most important work being done in the areas of chemistry and natural philosophy. A problem for Wilson in analyzing Cavendish's contributions to science is the fact that he was often reluctant to publish his research, so many of his important discoveries are found in his unpublished papers or were often not published until some time after he did his initial research. As we will see, this would be a major factor in the "Water Controversy." In terms of Cavendish's published research, all in the *Philosophical Transactions of the Royal Society*, Wilson sees four as being particularly important. One of the earliest was on the subject of electricity and entitled "An account of some Attempts to imitate the Effects of the Torpedo by Electricity," which was read to the Royal Society in 1775 and published the next year (G. Wilson 1851, 466–470). The torpedo is a ray capable of delivering an electric shock but there was some debate if the shock was really electric since it did not produce a spark, or if such animal electricity differed from other forms of electricity. Since torpedoes are only found in the Mediterranean, Cavendish constructed two "artificial torpedoes" out of wood, leather, pewter plates, and sheepskin attached to a Leyden jar. Using these devices, he conducted a number of experiments that proved that the phenomenon was truly electric and that it was the same type of electricity that was found in a Leyden jar. In the process of doing these experiments, Cavendish showed that electricity could act in a number of different ways and developed the important distinction between the intensity of electricity (what is now called voltage) and the quantity of electricity (what is now called amperage or current). He was also able to anticipate some elements of what became known as Ohm's Law by showing that a small quantity and a large intensity of electricity produced the same effect as a large quantity and a small intensity of electricity. Michael Faraday specifically commented on Cavendish's work and its importance for understanding electricity.

Wilson sees that two of Cavendish's major contributions to science were in the field of chemistry, and here he comments that Cavendish was in the tradition of such chemists as Robert Boyle, Robert Hooke, Stephen Hales, Joseph Black, and Joseph Priestley, who were followers of Bacon (G. Wilson 1851, 24, 35). Cavendish's two important contributions to chemistry arose from research that he began in 1777 or 1778 on *Experiments on Air* that were undertaken to discover the quantitative composition of the atmosphere (G. Wilson 1851, 31–49, 231–264). In the course of these experiments, Cavendish came to the conclusion that water was not an elemental substance but arose from oxygen and hydrogen (what he called inflammable air and dephlogisticated air). This would lead to the "Water Controversy" that we will discuss later. In his experiments leading to the composition of water, Cavendish also discovered that nitric acid, which had been thought to be an elemental substance, was also a compound of nitrogen and oxygen (what he called phlogisticated air and dephlogisticated air). Joseph Priestley and a friend had exploded gases, such as oxygen and hydrogen, in glass and metal vessels as a way to amuse friends and had noticed that often there was moisture left in the vessel that Priestley attributed to condensation of water that was in the gases. But when Cavendish heard of this, he decided that the water might be related to the disappearance of oxygen when hydrogen exploded in the air. During the summer of 1781, he began a detailed set of experiments, exploding various combinations of hydrogen and oxygen until with the right combination all of the hydrogen and oxygen disappeared, and an equal weight of water was left behind. He concluded that hydrogen and oxygen were being converted into water and, therefore, water was not an elemental substance but a compound of hydrogen and oxygen. In some of these experiments, he discovered that the water was acidic and upon further experimentation concluded that it was the result of some small amount of residual air in the container being mixed with the hydrogen and oxygen. Since the air was mostly nitrogen, this led him to the conclusion that nitric acid could be created through a combination of nitrogen and oxygen (although he still believed that nitric acid was a fundamental substance while nitrogen was a combination of nitric acid and hydrogen). In doing this experiment, Cavendish seemed to use a version of Bacon's principle of exclusion. After showing that a combination of pure oxygen and pure nitrogen produced nitric acid when combusted by a spark, he then showed that a vessel with either just pure oxygen or just pure nitrogen would not produce nitric acid. Wilson sees the discovery of the composition of water and of nitric acid as two of Cavendish's major chemical discoveries. Wilson does not credit Cavendish with the discovery of hydrogen since it had been identified by others but notes that he did the most thorough study of it and modern historians of chemistry credit Cavendish with its discovery (Ihde 1964, 747).

Wilson identifies Cavendish's fourth major scientific discovery as his 1798 experiment to determine the density of the earth, which led to him being labeled the "man who weighed the earth" (G. Wilson 1851, 471–474). In what came to be known as the "Cavendish experiment," which is still often repeated today in many science classes, a torsion balance, invented by John Mitchell, but improved by Cavendish is used to measure the force of gravity. In the experiment, a bar with two weights at each end is hung by a wire in the middle. Two larger weights are then brought near the weights at the end of the bar, and the gravitational force causes the two smaller weights to be attracted to the larger weights causing the bar to twist causing an oscillation. The rate of oscillation can be used to find the strength of the gravitational force that can then be used to find the density of the earth and the density can be used to find the entire weight of the earth. This data was useful in solving a number of astronomical and geological problems.

For Wilson, the main value of Cavendish's life seems to be his experiments. It should be noted that the modern biographers of Cavendish subtitled their book "The Experimental Life" (Jungnickel and McCormmach 1999). Wilson concludes his biographical section of his book stating that Cavendish "has enriched us all, by his lessons and his example, by his method of research and his great discoveries ..." (G. Wilson 1851, 190). Wilson presents that method in very Baconian terms when he says that for Cavendish, the universe seems to have

> consisted solely of a multitude of objects which could be weighed, numbered, and measured; and the vocation to which he considered himself called was, to weigh, number, and measure as many of those objects as his allotted three-score years and ten would permit.
>
> (G. Wilson 1851, 186)

As noted earlier, the vast majority of Wilson's "biography" of Cavendish has to do with the "water controversy," almost all of which took place after Cavendish had died. Wilson himself was one of the leading figures in the creation of the controversy that took place between the 1830s and 1850s (Miller 2004). The water controversy involved the question of who was the original discoverer that water was not a fundamental element, along with earth air and fire, but a combination of other elements. Essentially, three people had a claim to this discovery, although the water controversy centered mostly on two. One issue was what the word discovered meant. The famous French chemist Lavoisier was the first to describe the modern idea that water was composed of oxygen and hydrogen and, therefore, could be claimed to be the discoverer, but he was not the first to demonstrate that water was not a fundamental substance but a combination of other elements. Two Britons claimed to have made this discovery – Henry Cavendish and James Watt. Watt is more often thought of as the inventor of the steam engine and not a chemist, but while inventing the steam engine Watt collaborated with the chemist Joseph Black on issues concerning the latent heat of steam, and Watt maintained an interest in the chemistry of water in order to improve his steam engine. Most of the water controversy involved the claims of Cavendish and Watt to be the discoverer that water was a compound substance. Two things add to the confusion. First, Cavendish, who was often slow in publishing the results of his research, appears to have conducted experiments concerning the composition of water in the summer of 1781 but his paper was not read before the Royal Society until 1784 (G. Wilson 1851, 55–62). During the intervening period, Cavendish privately communicated the results of his experiment to Joseph Priestley and to Lavoisier, and Priestley subsequently communicated the results to Watt, without mentioning Cavendish. Watt then began his own set of experiments which showed that water was a compound and communicated his results back to Priestley. When Watt heard of Cavendish's public announcement, Watt, encouraged by some other individuals, came to the assumption that Cavendish had seen Watt's letter to Priestley and had plagiarized his results. While all of this led to bad feelings which show up in Watt's and Cavendish's private correspondence with friends, it never broke out in public. According to Wilson, Watt never publicly accused Cavendish of plagiarism and during a visit Cavendish made to Watt in Birmingham in 1785 their meeting was a friendly one (G. Wilson 1851, 76, 162). A second element adding to some of the confusion concerning the discovery of the composition of water was the fact that both Cavendish and Watt still embraced the doctrine of phlogiston, which we discussed in the first chapter. Therefore, unlike Lavoisier, who was using modern-day terminologies, such as oxygen and hydrogen, Cavendish and Watt were using somewhat vague terms, such as inflammable air or dephlogisticated air, and while Cavendish seems to have associated hydrogen with phlogiston, it is not clear that Watt did so, making it hard to decide if he discovered that water was a combination of hydrogen and oxygen (G. Wilson 1851, 157).

The actual water controversy did not begin until the 1830s after Cavendish, Watt, and Lavoisier had died and was partially stoked by the publication of Watt's private correspondence in 1846 (Miller 2004, 207–208, 236–241; Muirhead 1846). The nineteenth-century version of the water controversy effectively began in 1834, when the chemist François Arago, Perpetual Secretary of the Académie des Sciences, read his *éloge* of James Watt, who had been a member of the Académie (Miller 2004, chap. 6). It was translated into English and published in 1839. Much of the material for the *éloge* had been provided by James Watt, Jr., Watt's only surviving son, who wanted to uphold his father's claim as being the discoverer of the composition of water. David Philip Miller argues that Arago had his own reasons to build up Watt since he wanted to see science more closely associated with technology (Miller 2004, 7). Given these circumstances, Arago's *éloge* was very critical of Cavendish, claiming the "crime of Cavendish" in plagiarizing Watt's work (Miller 2004, 126–127).

British reaction came in 1839, the same year of the English translation of Arago's work (Miller 2004, chap. 7). In that year, the British Association for the Advancement of Science met in Birmingham. While the city was closely associated with the work of Watt, William Vernon Harcourt, President of the Association, took the opportunity to respond to Arago's attack and put forward a detailed defense of Cavendish claiming that Cavendish's experiments preceded Watt's and that Cavendish's train of reasoning was precise while Watt's was vague. Wilson may have become interested in Cavendish and the water controversy after attending the British Association meeting in Birmingham and he may have become further interested through Thomas Graham, who was his former mentor, when Graham proposed a project in 1846 to publish books on chemical subjects and named it the Cavendish Society and it was the Cavendish Society that commissioned Wilson to write his *Life of Cavendish* (Miller 2004, 206). Wilson also had connections with members of the Watt camp, including James Patrick Muirhead, editor of the Watt *Correspondence*, and Lord Francis Jeffrey, who had been friends with Watt and Watt, Jr. Both Muirhead and Jeffrey convinced Wilson to write a review of the Watt *Correspondence* for the *Edinburgh Review,* which was being edited at the time by Jeffrey, but Jeffrey thought that Wilson's article did not give enough credit to Watt and the review was then written by Jeffrey (Miller 2004, 207–208). Miller argues that the rejected review of the Watt *Correspondence* became the beginning point of Wilson's *Life of Cavendish*.

As noted earlier, almost 300 pages of Wilson's *Life of Cavendish* is devoted to the water controversy. In these pages, Wilson provides an incredibly detailed analysis of all aspects of the controversy, including published and unpublished researches, letters, and notebooks of all of the participants, such as Cavendish, Watt, Priestley, Lavoisier, and others involved in the original experiments. In addition, Wilson provides a detailed analysis of all of the arguments of the participants involved in the nineteenth-century debate, such as Harcourt, Muirhead, Watt, Jr., Jeffrey, and Lord Brougham. It is almost as if Wilson is presenting evidence for a legal case. In the end, Wilson concludes that Cavendish should be considered the discoverer of the composition of water, and that Watt should be given credit as an original theorist on the composition of water and a person who played an important role in disseminating the true theory of the composition of water (G. Wilson 1851, 157, 432–445). He goes on to credit Lavoisier with simplifying Cavendish's conclusions and explaining them in terms of modern chemical theory. His support of Cavendish over Watt seems to be based on their styles of research. Although Cavendish did not explicitly state the composition of water before his public paper in 1783, Miller interprets Wilson arguing that Cavendish's entire research program from 1781 to 1783 "only made sense if he had drawn conclusions about the composition of water at a stage prior to Watt" (Miller 2004, 211). On the other hand, Wilson argues that Watt's research program was aimed more at the erroneous idea that steam could be changed into atmospheric air through heat and that he would have never considered that water was a compound of oxygen and hydrogen unless he had heard of Priestley's repetition of Cavendish's 1781 experiment (Miller 2004, 211–212; G. Wilson 1851, 437). Wilson also accuses the supporters of Watt of having a naïve conception of the nature of an experiment. He says they

> would, apparently, have us believe, that [an experiment] is only the handling of certain pieces of apparatus, and the unreflecting observation of certain phenomena, without any hypothesis as to their cause, or any conclusion as to their significance.
>
> (G. Wilson 1851, 80)

This seems to make clear that Wilson has gone beyond the naïve version of Baconianism and like the nineteenth-century Common Sense philosophers accepted that hypothesis is part of the experimental process, otherwise Cavendish's research program from 1781 to 1783 would not make any sense. For Wilson, an important part of the water controversy was the accusation of plagiarism against Cavendish which Wilson seems to have found offensive (G. Wilson 1851, 407–432). In fact, one of his main arguments in favor of Cavendish as the discoverer of the composition of water was that he, "a highly honourable, upright, and modest man claimed the conclusion as his own" (G. Wilson 1851, 435).

Miller makes another argument why Wilson and many other British chemists took the side of Cavendish in the water controversy (Miller 2004, 202–214). He notes that Wilson's book was published in 1851, the same year as the Great Exhibition that was praising the ways in which science could benefit Britain's industrial development. Throughout the 1840s, British chemistry was becoming more professionalized with the founding of the Chemical Society of London, the Pharmaceutical Society, and the Royal College of Chemistry. Many of the new professional chemists saw Cavendish as a model given his careful, experimental, disinterested, and quantitative approach that was reflected in the group of London chemists naming their new publishing venture the Cavendish Society. On the other hand, Watt's approach did not seem to reflect the values of the "new chemists." The importance of a new professionalized approach to chemistry, personified by Cavendish, was reflected in a number of addresses given in association with the Great Exhibition. Miller quotes Whewell as saying: "The great chemical manufactories which have sprung up at Liverpool, at Newcastle, and Glasgow, owe their existence entirely to a profound and scientific knowledge of chemistry" (Miller 2004, 212). Lyon Playfair, a friend of Wilson's who had read and referred to his *Life of Cavendish*, argued that industry would benefit from the development of abstract science, such as chemistry, and went on to say: "The impersonification of abstract science was Cavendish ...His discovery of the composition of water has given to Industry a vitality and an intelligence, the effects of which would be difficult to exaggerate" (Miller 2004, 213). Wilson seems to support this role for Cavendish. Near the end of the biographical section of his book, Wilson discusses the importance of Cavendish's discovery of the composition of nitric acid and quotes Joseph Black as saying: "This discovery by Mr. Cavendish is one of the most important in the whole science of chemistry" (G. Wilson 1851, 263). Wilson goes on to argue that besides explaining the source of atmospheric nitric acid, Cavendish's discovery also explained nitrates in the soil. Some years later, this would become an important discovery when Justus von Liebig began understanding the role of nitrates and plant physiology and uses that knowledge to develop chemical fertilizers. Therefore, Wilson's *Life of Cavendish* and his work on the water controversy need to be seen as representing important changes that were taking place in Victorian science.

John Reid

Wilson's last scientific biography was the *Life of Dr. John Reid* that was published in 1852 (G. Wilson 1852). Although he began a biography of his friend Edward Forbes, Wilson died after completing only a few chapters and most of the work was written by Archibald Geikie, one of Scotland's leading geologists (G. Wilson and Geikie 1861). John Reid (1809–1849) was not as famous as the subjects of Wilson's other biographies, but he was a well-known physiologist. Wilson may have chosen him for a biography since they had been friends and their lives paralleled each other's in many ways. Reid had graduated from the University of Edinburgh Medical School in 1830 just a few years before Wilson and overlapped with him teaching at the School of Anatomy in Old Surgeons' Hall, before becoming Chandos Professor of Medicine and Anatomy at the University of St. Andrews. Wilson may have been especially attracted to write a biography of Reid since they both went through a very similar religious conversion experience. At the height of Reid's career when he was only thirty-eight, he began suffering from symptoms that would be diagnosed as cancer of the tongue, which had a very low survival rate. While he had been brought up in a Christian household, he was only nominally a Christian for most of his life (G. Wilson 1851, 200–202). Like Wilson, who had a conversion experience while facing death, soon after his diagnosis Reid had a conversion experience and became a devoted evangelical Christian. Wilson dedicates almost one-third of his biography to discussing Reid's religious beliefs and feelings which led to some criticism of the work. John Goodsir, part of the Forbes Brotherhood, was concerned when he learned that Wilson was going to write the biography of Forbes because he found Wilson treated Reid's life more from a theological point of view rather than focusing on Reid as a physiologist (Goodsir and Lonsdale 1868, 167). Although Wilson does spend much time on Reid's religious conversion, the first part of the book presents Reid's scientific accomplishments and focuses on his methodology. Reid had studied anatomy with Robert Knox, the famous transcendental anatomist, and after a brief period of study in Paris after obtaining his M.D. he returned to Edinburgh to become Demonstrator of Anatomy under Knox at the Extra-Academical School of Anatomy in Old Surgeons' Hall. Although he spent several years with Knox, Reid was never converted to Knox's transcendental anatomy choosing to follow a much more strictly Baconian approach (G. Wilson 1852, 40–41, 304). While in Edinburgh, Reid also worked as assistant physician in the Royal Infirmary and later as a lecturer at the Extra-Academical Medical School. He later was appointed a Pathologist to the Royal Infirmary and later made Superintendent.

Since St. Andrews did not have a hospital, most of Reid's scientific work in human physiology was done while he was still in Edinburgh. At St. Andrews, his research was limited to natural history, particularly the physiology of sea creatures. Reid's most famous research was on the functioning of a set of nerves, including the glasso-pharyngeal, which influences the tongue and pharynx, and the pneumogastric nerve that influences the heart, lung, and digestive tract. Since these nerves control many of the important organs of the body, understanding their function and how they can be affected by disease was important for medicine. Doing numerous and detailed experiments, many on animals, Reid was able to shed light on how these nerves functioned and cleared up previous misunderstandings (G. Wilson 1852, 136–147). He was able to show that the glosso-pharyngeal nerve was a nerve of sensation and that it was excito-motory and taste depended on it. Reid's work on the pneumogastric nerve was said by the English physiologist, William Benjamin Carpenter, to be "the most laborious and scrutinizing investigation of any single nerve, every carried through by any single physiologist" (G. Wilson 1852, 142). Reid was the first to show that the nerve functioned in an automatic fashion, controlling coughing and other functions. He was also able to show that the nerve's role in digestion disproved a popular notion that digestion was a purely electrical phenomenon by showing that digestion could still take place even if the nerve was severed. The British physiologist J. Hughes Bennett said that Reid's research "contains more original matter and sound physiology, than will be found in any work that has issued from the British Press for many years" (G. Wilson 1852, 110).

Wilson makes the important argument that Reid's success in his scientific research was the result of strictly adhering to the Baconian method. He quotes Reid as saying:

> In entering upon this investigation, I had no favourite theory to defend, stood committed to no preconceived notions, nor shackled by any slavish defense to authorities, but was ready and willing to give up any of my former opinions as they appeared to be inconsistent with the phenomena which presented themselves ....
>
> (G. Wilson 1852, 139–140)

Wilson further outlines Reid's method saying that he "began an inquiry by carefully studying all that had already been written on the subject" and "His method of research was a purely inductive as it could well be" (G. Wilson 1852, 148–150). Wilson goes on to say:

> In intellectual character, he was such a man as Bacon would have given his right hand to, and would have heartily trusted to work out and illustrate in practice the precepts of the Novum Organon [*sic*]. Seldom has a purer specimen of an inductive reasoner been seen than was supplied by John Reid.
>
> (G. Wilson 1852, 304)

Wilson also uses Reid's experimental method to raise important questions concerning the morality of certain types of research. While Reid's experiments followed the Baconian method, they also required the infliction of pain and suffering on animals (G. Wilson 1852, 152–164). Wilson, a devout animal lover, raises the question of how far one can inflict suffering on animals in order to benefit humans. He notes that Reid was gravely concerned about the issue but goes on to say that the issues are not just a problem for researchers, but that everyone benefiting from that research must address the issues. Wilson argues for two principles in doing animal research: "1. Pain should be inflicted only to serve a useful end. 2. Pain should be inflicted as sparingly as possible" (G. Wilson 1852, 156). He goes on to note that Reid followed these principles in his experiments. Wilson goes on to conclude that the problem of animal suffering in experiments is not just a moral problem but can also become an intellectual problem since it can undermine the intellectual reverence for life.

The Electric Telegraph

Wilson's biographical works seem to all address more or less explicitly, the methodology of science, especially a Baconian or inductive approach, but he also wrote other works that often only implicitly touched on the history of science and the methodology of science. In 1852, Wilson published a book entitled *Electricity and the Electric Telegraph* which for unclear reasons also included his paper "The Chemistry of the Stars," which we discussed in the last chapter (G. Wilson 1859). The telegraph article was originally written at the request of Lord Jeffrey for the *Edinburgh Review* in 1849 which he said was an "admirable paper, giving a luminous account of the invention [the telegraph]" (J. Wilson 1866, 199). A review of the first edition says:

> If any one is destined to open up a royal road to science, it is Dr. Wilson.... It is seldom that we find a man so eminent in science retaining all the warmth and freshness of humanity about him. He clothes every subject he touches with the bright hues of fancy and the warm sympathies of a human heart.
>
> (J. Wilson 1866, 199–200)

The book is mostly a straightforward description of the history of electricity, the history of magnetism, the discovery by Oersted and Faraday of the connection between the two, and a discussion of the invention of the first telegraphy. There is little, if any, discussion of the ideology of science and technology or of their methodologies except for the implication that the interrelationship of electricity and magnetism reflects a certain unity and the example provided by the telegraph that scientific discoveries can have practical benefits.

The Five Gateways of Knowledge

Probably, Wilson's most famous work was a book entitled *The Five Gateways of Knowledge* (G. Wilson 1857). The book began as a Sunday School lecture in Leith later presented as two lectures to the Philosophical Institution in Edinburgh and finally published as a book (J. Wilson 1866, 301–302). The book was very well received and within five years had sold 8,000 copies. He notes that he received "Hearty" responses from Charles Dickens, John Ruskin, and William Wordsworth, among others (J. Wilson 1866, 303). Dickens seemed especially impressed with the book, writing Wilson a letter on November 4, 1856, saying:

> I am truly honoured by your kind remembrance in having sent me the Five Gateways of Knowledge, and I beg to assure you that I have read that charming little book with the highest interest and gratification. Wise, elegant, eloquent, and perfectly unaffected, it has delighted me. I should find it very difficult indeed to tell you, to my own satisfaction, how unusually it has pleased me.

(Hollington 2013, 29)

The book begins with a quote from John Bunyan concerning the isolated city of Mansoul that could only be entered by five gates – the ear-gate, the eye-gate, the mouth-gate and nose-gate, and the feel-gate. Wilson argues that like the fortress city of Mansoul, the five senses are "at once loopholes through which the spirit gazes out upon the world, and the world gazes in upon the spirit..." (G. Wilson 1857, 5–6). He goes on to say that while the senses play a role in the purely animal wants of the body, more importantly, they help to cultivate the intellect as well as the esthetical perception of beauty and that our moral instincts flow "from the triple Corporeal, Intellectual, and Æsthetical function which is exercised by each sense" (G. Wilson 1857, 8). Although he does not specifically mention Common Sense, the relationship between the senses and the moral was a foundational element of that philosophy.

What follows are five chapters, each for one of the sense organs, that first explain the physiology of each sense organ in very simple everyday terms. For example, he compares the shape of the eye to an acorn, or eggcup, or wineglass, the eyelids are compared to outside shutters which keep glass windows clean, the iris is compared to Venetian blinds or a daisy that can open and close, letting in more or less light, the lens is compared to a magnifying glass, and the retina as a curtain or a flower cup that ends in a stem. But Wilson goes beyond a purely physiological description of the function of the eye and argues that it does not function in a purely sensational or materialistic way. The senses created by the eye are not a simple mirror of the reality of the external world but those sensations, what he calls shadows, must be interpreted through interactions with the soul (G. Wilson 1857, 24–27). Again, although he does not refer to philosophy, the idea that we can only perceive objects by way of our senses and not directly reflects elements of Kantian philosophy that may have influenced him through Forbes's Brotherhood who were reading Coleridge's interpretation of Kant. The idea that sensations were shaped by some inherent element of the mind, such as Kant's categories or the Common Sense idea of instincts, is reflected by Wilson's statement that: "the two most conflicting of unlike existences, the dead world of matter and man's immortal soul, hold their twilight interviews, and make revelations to each other" (G. Wilson 1857, 28).

The idea that the senses are not simply passive but play an active role, not just in perception, but also in esthetical and moral understanding, is something Wilson repeats these ideas in each of his chapters. In discussing the eye, he says it "was intended by its Maker to be educated" and quotes Thomas Carlyle as saying: "The eye sees what it brings the power to see" (G. Wilson 1857, 32–33). The fact that he repeats the idea that each one of the senses can, and must, be educated seems to be one of the main messages and purposes of the book. As might be expected from Wilson, another purpose may have been to establish a connection between the senses and religious ideas. Not every chapter makes this connection, but in the chapter on the eye, he notes one of its roles is "to show us the beauty of Nature, and teach us the wisdom of God" (G. Wilson 1857, 27). He notes the role of the ear in sacred music and says of Handel's Messiah: "From these words we learn that the summons to the life to come will be addressed first to the Ear" (G. Wilson 1857, 70). Finally, Wilson connects the nose to incenses which he claims to be of divine origin and says that King David saw incense as having a greater reality than prayer and quotes him saying: "Let my prayer be set forth before thee as incense (Ps. cxli.2)" (G. Wilson 1857, 107). So while appearing to be a purely popular book on the five senses, Wilson provides a deeper message about the active role of the senses in our perception of both nature and God.

Color Blindness

Although he does not specifically mention Bacon, Wilson certainly was following both the Baconian method and the Baconian ideology in his most famous scientific work, *Researches on Colour Blindness* (G. Wilson 1855a). Wilson may have been attracted to the subject of color blindness from his earlier 1845 biographical sketch of John Dalton (G. Wilson 1862a, 304–364). In that sketch, Wilson notes that Dalton's first paper read to the Manchester Literary and Philosophical Society in 1794 was a description of his own color blindness and his discovery of the condition in other people, such as the philosopher Dugald Stewart (G. Wilson 1862a, 335, 358). As a result, on the Continent, color blindness became known as Daltonism and people with the affliction were referred to as Daltonians. Wilson adds an appendix to his book on *Colour-Blindness*, in which he argues that the British find such terms offensive (G. Wilson 1855c, 161–162). Wilson's book began as a series of papers published between 1853 and 1854 in the *Edinburgh Monthly Journal of Medical Science*, but since the medical journal wanted only the discussion of medical issues and not practical ones, Wilson added to his book a supplement on the dangers that color blindness could hold for railway and marine-colored signals possibly reflecting the Baconian ideology that knowledge should be useful (G. Wilson 1855a, iii–v; G. Wilson 1855b, 127–152).

Wilson's research on color blindness not only summarized the existing knowledge on the subject but he also did detailed studies and interviews with more than 1,000 people who had the condition, including students, soldiers, and hospital patients (G. Wilson 1855a, 68–77). This again may have reflected the Baconian methodology of developing histories and then collecting a large number of facts. From his research, he was able to conclude that about two percent of the population exhibited the type of color blindness that Dalton had and about five percent of the population had some type of color blindness. This is somewhat close to the modern estimate of eight percent of males having color blindness. He was not able to find any females with color blindness, but this also fits with the modern idea that females are only one-eighth as likely to have color blindness. Through his studies, he was able to conclude that a significant number of cases of color blindness arose from the inability of a patient to be able to perceive red light (G. Wilson 1855a, 53–68). Wilson hoped to find the cause of color blindness and investigated a number of theories such as the role of the yellow spot on the retina and the color of the choroid coat playing a role in color blindness and even speculates on the phrenological theories of George Combe as a way of understanding it (G. Wilson 1855a, 83–108). He also came to recognize that color blindness was hereditary and could be passed from either parent, but it affected male offspring more often than female (G. Wilson 1855a, 130). While none of these theories ultimately explained color blindness, he was able to suggest some practical ways to deal with the problem. First, he noticed that the problem was less pronounced under artificial light, such as gaslight and second, he discovered that wearing glasses with a yellow or orange tint could also alleviate the condition (G. Wilson 1855a, 118–123).

The most important impact of Wilson's research on color blindness was in developing new safety standards for the railway and shipping industries reflecting a practical application of knowledge. Ever since 1830, railways had been using a system of colored lights to control trains and in 1848 in order to prevent collisions, the British Admiralty began to require all ocean-going steamships have a green light on the starboard side and a red light on the port side and a white light on the mast (London Gazette July 11, 1848). In his "Supplement: On Railway and Ship Signals in Relation to Colour-Blindness," Wilson points out the dangers of having workers with color blindness on ships and railways, especially since the system of lights often relied on the colors red and green (sometimes yellow) which were a particular problem for people who were color-blind (G. Wilson 1855b, 127–159). Wilson makes some suggestions to lessen the problem. One was to always associate a color with a distinctive shape or placing flags and semaphores at distinctive angles. But for people with normal vision, color is of great advantage in distinguishing an object. He notes that red and green lights coming from artificial lights and used at night might be less of a problem than reflected colored signals used during the day. Since often other colors could also be confused, substituting them for red and green would probably not solve the problem and using one set of colors during the day and another set at night would be confusing. His best solution was to make sure that those workers dealing with safety signals be tested for color blindness, both at different distances and under different levels of illumination. Wilson notes that James Clerk Maxwell in a letter to him and included in the book describes how a disc rotated like a top and divided into sections with different colors can be used as a simple test for color blindness (Maxwell 1855, 153–159). Wilson's work on color blindness received a great deal of praise from other scientists, such as Maxwell, John Tyndall, John H. Gladstone, and David Brewster. In 1857, at an ophthalmological congress held in Brussels, a report by White Cooper said the following:

> Though I have abstained from making special reference to books, I cannot pass over the admirable and original work on chromate-pseudopsis, or colour-blindness, by Dr. George Wilson, of Edinburgh. For its acuteness and originality, this volume deserves the highest praise.
>
> (J. Wilson 1866, 187)

Wilson wrote a number of other papers concerned with technology which will be addressed in the next two chapters. But it was essentially in his scientific biographies that he addressed issues concerning the methodology of science and put forward the idea of a methodology of science that drew from both Baconian philosophy and Common Sense philosophy. As we have seen, both of these philosophies emphasized the role of the senses in gaining knowledge, but Common Sense also argued for an innate capacity of the mind to shape these sense impressions. Also, as we have seen, by the nineteenth century, both followers of Bacon and followers of Common Sense were much more open to seeing some value of hypotheses in the methodology of science.

References

Bacon, Francis. 1937. *Essays, Advancement of Learning, New Atlantis, and Other Pieces*, edited by Richard Foster Jones. New York: Odyssey Press.

Balfour, John. 1860. *Biographical Sketch of the Late George Wilson, M.D.* Edinburgh: Murray and Gibb.

Briggs, John C. 1989. *Francis Bacon and the Rhetoric of Nature*. Cambridge, MA: Harvard University Press.

Brougham, Lord Henry. 1845. *Lives of Men of Letters and Science, Who Flourished in the Time of George III*. London: Charles Knight and Co.

Cairns, John. 1860. "The Late George Wilson of Edinburgh." *Macmillan's Magazine* 1: 199–203.

Goodsir, John, and Henry Lonsdale. 1868. *Anatomical Memoirs of the Late John Goodsir*, vol. 1, edited by William Turner. Edinburgh: Adam and Charles Black.

Henry, William Charles. 1854. *Memoirs of the Life and Scientific Researches of John Dalton*. London: Cavendish Society.

Hollington, Michael. 2013. "Dickens, the City, and the Five Senses." *Journal of the Australasian Universities Language and Literature Association* 2010: 29–38.

Ihde, Aaron J. 1964. *The Development of Modern Chemistry*. New York: Harper & Row.

Jungnickel, Christa, and Russell McCormmach. 1999. *Cavendish: The Experimental Life*. Lewisburg, PA: Bucknell University Press.

London Gazette. 1848. 20876 (July 11, 1848); 2606.

Martin, Julian. 1992. *Francis Bacon, the State, and the Reform of Natural Philosophy*. Cambridge: Cambridge University Press.

Maxwell, James Clerk. 1855. "On the Theory of Colous in Relations to Colour-Blindness in a Letter to Dr. G. Wilson." In *Researches on Colour-Blindness*, edited by George Wilson, 153–159. Edinburgh: Sutherland and Knox.

Miller, David Phillip. 2004. *Discovering Water: James Watt, Henry Cavendish and the Nineteenth-Century "Water Controversy"*. London: Routledge.

Muirhead, James Patrick, ed. 1846. *The Correspondence of the Late James Watt on His Discoveries of the Composition of Water*. London: John Murray.

Olson, Richard. 1975. *Scottish Philosophy and British Physics 1750–1880*. Princeton, NJ: Princeton University Press.

Provincial Medical and Surgical Society. 1846. "Announcement." *Provincial Medical and Surgical Society Journal* 10: 446–447.

Rehbock, Phillip F. 1983. *The Philosophical Naturalists: Themes in Early Nineteenth-Century British Biology*. Madison: University of Wisconsin Press.

Rossi, Paolo. 1978. *Francis Bacon: From Magic to Science*. Chicago: University of Chicago Press.

Sacks, Oliver. 2001. "Henry Cavendish: An Early Case of Asperger's Syndrome?" *Neurology* 57: 1347.

Shapin, Steven, and Simon Schaffer. 1985. *Levithan and the Air-Pump: Hobbes, Boyle, and the Experimental Life*. Princeton, NJ: Princeton University Press.

Usselman, Melvyn C. 2015. *Pure Intelligence: The Life of William Hyde Wollaston*. Chicago: University of Chicago Press.

Wayling, H.G. 1927. "A Short Biography of William Hyde Wollaston, M.D. F.R.S." *Science Progress in the Twentieth Century (1919–1933)* 22: 81–95.

Webster, Charles. 1974. *The Great Instauration: Science, Medicine, and Reform, 1626–1680*. London: Duckworth.

Whewell, William. 1837. *History of the Inductive Sciences, from the Earliest to the Present Times*, 3 vols. London: J.W. Parker.

Whewell, William. 1840. *Philosophy of the Inductive Sciences, Founded on Their History*, 2 vols. London: J.W. Parker.

Wilson, George. 1846. "On the Alleged Antagonism between Poetry and Chemistry." *The Torch: Journal of Literature, Science, and the Arts* 1: 13–16.

Wilson, George. 1849. *On the Early History of the Air-Pump in England*. Edinburgh: Neal and Co.

Wilson, George. 1851. *The Life of Henry Cavendish, Including Abstracts of His More Important Scientific Papers, and a Critical Inquire into the Claims of All the Alleged Discoveries of the Composition of Water*. London: The Cavendish Society.

Wilson, George. 1852. *Life of Dr. John Reid*. Edinburgh: Sutherland and Knox.

Wilson, George 1855a. *Researches on Colour-Blindness*. Edinburgh: Sutherland and Knox.

Wilson, George. 1855b. "On Railway and Ship Signals in Relation to Colour-Blindness." In *Researches on Colour-Blindness*, edited by George Wilson, 153–159. Edinburgh: Sutherland and Knox.

Wilson, George. 1855c. "On the Terms Daltonism and Daltonians." In *Researches on Colour-Blindness*, edited by George Wilson, 161–162. Edinburgh: Sutherland and Knox.

Wilson, George. 1857. *The Five Gateways of Knowledge*. Philadelphia: Parry and McMillan.

Wilson, George. 1859. *Electricity and the Electric Telegraph Together with the Chemistry of the Stars: An Argument Touching the Stars and Their Inhabitants*. London: Longman, Brown, Green, Longmans & Roberts.

Wilson, George, and Archibald Geikie. 1861. *Memoir of Edward Forbes, F.R.S.* London: Macmillan and Co.

Wilson, George. 1862a. *Counsels of an Invalid: Letters on Religious Subjects*. London: Macmillan and Company.

Wilson, George. 1862b. "Life and Discoveries of Dalton." In *Religio Chemici: Essays*, edited by George Wilson, 304–364. London: Macmillan and Company.

Wilson, George. 1862c. "Robert Boyle." In *Religio Chemici: Essays*, edited by George Wilson, 165–252. Macmillan and Company.

Wilson, George. 1862d. "Wollaston." In *Religio Chemici: Essays*, edited by George Wilson, 253–303. London: Macmillan and Company.

Wilson, Jessie Aitken. 1860. *Memoir of George Wilson*. London: Macmillan and Co.

Wilson, Jessie Aitken. 1866. *Memoir of George Wilson*, new edition. London: Macmillan and Co.

Yao, Richard. 1985. "An Idol of the Market-Place: Baconianism in Nineteenth Century Britain." *History of Science* 23: 251–301.

Zagorin, Perez. 1998. *Francis Bacon*. Princeton, NJ: Princeton University Press.

5 The Eye and the Hand

Wilson's Biological View of Technology

Regius Professor of Technology

In August of 1855, George Wilson received word that Queen Victoria had appointed him Regius Professor of Technology at the University of Edinburgh. He received it in an unusual way. According to his sister, they were on a brief holiday in the town of Melrose and when George could not find a newspaper, he rode to the next town to buy a copy of the *Scotsman*. While reading it on the way home, he said to his sister: "This is decidedly worth a penny, read that" (J. Wilson 1866, 276). What she read was a notice that Wilson had been appointed Professor of Technology. He would receive the official document later by mail. A few months earlier in February, he had been appointed Director of the new Industrial Museum of Scotland. Wilson's sister argues that the chair

> was suggested first by the professors in Edinburgh University, to whom it seemed more advisable to have the director of the New Museum amenable to their laws, than to have in him one who might set up rival claims as a public teacher, with a salary from Government, and valuable museums at his disposal.
>
> (J. Wilson 1866, 276)

R.G.W. Anderson confirms this with a quote from Lyon Playfair, who arranged the joint appointment, saying:

> The disadvantage of having [Wilson's classes] unconnected with the University is that the lecturer might come into collision with the courses on Chemistry, Mechanics, and Natural History in the college, and thus a feeling of hostility might arise between the Museum and the University, when it is so desirable to have entire co-operation and identity of interests.
>
> (Anderson 1992, 177)

There was already a precedent for such a joint appointment. Also, in 1855, the University's collection of natural history was transferred to the Board

DOI: 10.4324/9781003212218-5

of Trade which had also just established the Industrial Museum (Yanni 1999, 92). The intention was to create a newly named Natural History Museum that would share a newly constructed building with the Industrial Museum (Swinney 2016, 170). Since the Keeper of the new Natural History Museum, George Allman, was also Professor of Natural History at the University, it made sense to have Wilson hold an appointment at the University.

The chair did not increase Wilson's salary but in his typical manner, he accepted it because he saw it as a religious calling. In a letter to John Hall Gladstone, a fellow chemist he met while working in Thomas Graham's laboratory in London, Wilson said: "As for the Chair, I trust and pray that it will increase my power to serve my blessed Lord and Master" (J. Wilson 1866, 277). If there were concerns among the other faculty about the chair, the public seemed very supportive. His sister notes that one of the periodicals at the time wrote:

> The formation of the Industrial Museum would in fact have been a matter of comparatively little importance to the community generally, had not this appointment [that of the new Chair] been made; and had the Government sought through the length and breadth of the land for a person fitted for carrying out the object contemplated by it, they would not readily have found one so well qualified as Dr. George Wilson.
>
> (J. Wilson 1866, 277–278)

At first glance, it appears unusual that a person with training in medicine and chemistry would be appointed to a chair in technology and to direct an industrial museum, but this can be explained by several factors. First, it was Wilson's old friend Lyon Playfair who was in charge of the Department of Science and Art at the Board of Trade and the person behind the creation of the Industrial Museum, and it seems from the earlier quote that Playfair played a role in establishing Wilson's joint appointment. Second, chemistry at the time was often seen as having closer ties to industrial development than to what we would call today pure science. In a biographical sketch, John Gladstone writes that even before his appointment to the Industrial Museum, "he had, in his laboratory practice, been led to investigate several of the chemical arts," and specifically he mentions work on the theory of bleaching (Gladstone 1860, 515). In his biographical sketch of Wilson, John Balfour, Professor of Botany at the University, said that Wilson was brought to the attention of the Department of Science and Art because of the "attention which he devoted to Economical Science, and to the applications of Chemistry" (Balfour 1860, 14). Finally, there is evidence that Wilson's work on color blindness, discussed in an earlier chapter, brought him to the attention of industrialists in the railway and shipping industries because of its practical application at a time when both industries were beginning to use colored lights as safety devices. In an article published in the *North British*

Review in 1856, the writer, later thought to be David Brewster, the well-known Scottish scientist, said:

> We have no doubt the researches [on color blindness] which it contains and their practical relation to the safety of ships and railway trains, which he was the first to point out, were among the grounds of his appointment to the Chair of Technology, or Industrial Art, which has recently been founded by the Crown in the University of Edinburgh.
>
> (J. Wilson 1866, 278)

But, this new appointment as Regius Professor of Technology caused Wilson to have to address other issues and problems. Wilson may have been aware that problems had arisen at Glasgow University over the creation of another Regius Chair (Channell 1982, 43–44). In 1840, the Crown created a Regius Professorship in Civil Engineering and Mechanics at Glasgow University in order to honor James Watt who had been an instrument maker at the University. Lewis D.B. Gordon, a relatively unknown engineer, was appointed to the chair and faced immediate problems. First, a number of the faculty did not think that engineering was a suitable university subject, and second, the senate requested a meeting "in reference to his not encroaching on or interfering with any of the present classes in the University" (Fortuna Domus 1952, 336–337). Gordon noted that he "was met with much jealousy by the Professors of Natural Philosophy and Mathematics" (Constable 1877, 227), which resulted in his not being able to find a classroom until the Lord Advocate intervened in 1841 (Fortuna Domus 1952, 337). Given his tenuous position in the university, Gordon was forced to suspend lectures during some sessions, due to low attendance, and he eventually resigned from the position in 1855 and was replaced by W.J.M. Rankine who met with more success in the position.

What Is Technology?

Wilson was facing many of the same problems at Edinburgh when he accepted the chair, and he faced them directly in his *What is Technology? An Inaugural Lecture Delivered in the University of Edinburgh on 7 November 1855* (G. Wilson 1855). Wilson began his lecture by noting that most of his audience was probably not familiar with the term "technology." There do not seem to be any records of why the chair was labeled Professor of Technology. Wilson noted that the term technology had been widely used in Germany and one might speculate that Prince Albert, who was both German and a great supporter of industrialization, and Playfair, who studied in Germany, may have played roles in naming the chair. In his book *Technology: Critical History of a Concept*, Eric Schatzberg writes that the term technology has a long and confusing history (Schatzberg 2018). He notes it was often used in at least three different ways: as meaning industrial arts, or meaning applied science, or meaning technique (Schatzberg 2018, 13).

In his *Inaugural Lecture*, Wilson discusses some of the confusing aspects of the term technology, including that its foreign origins make it "so unfamiliar to English ears, and so inexpressive to English minds" (G. Wilson 1855, 3). Possibly referring to its German origins, he later says: "Its sound is harsh and unwelcome ..." (G. Wilson 1855, 25). In a lecture a year later, he offers his audience the possibility of the term "industrial science," if they find the term technology too foreign (G. Wilson 1856b, 3). Wilson notes that the word technology appears to have been first used in 1772 by Johann Beckmann who in lectures at the University of Göttingen on the industrial arts used the term technology to refer to the science of these arts (G. Wilson 1855, 3). After discussing the Greek roots of *techne* and *logos*, Wilson notes that if taken literally, technology "signifies the Science of the Arts, or a Discourse or Dissertation on these" (G. Wilson 1855, 4). Wilson settles on defining technology as the "Science, or Doctrine, or Philosophy, or Theory of the Arts" (G. Wilson 1855, 4). He makes the point that the object of technology "is not Art itself, *i.e.* the *practice* of Art, but the principles which guide or underlie Art..." (G. Wilson 1855, 4). This focus on the "principles which guide or underlie Art" may be rooted in the transcendental natural history that Wilson was taught as a student of Robert Jameson and that was practiced by his close friend Edward Forbes, who before his early death had returned to the University of Edinburgh as Professor of Natural History. As discussed in earlier chapters, transcendental natural history sought to discover the basic ideal patterns and laws that existed behind or beyond the world of nature (Rehbock 1983, 7–10). Wilson's definition of technology seems to simply shift the focus of ideal patterns and laws from the natural world to the industrial world. This idea may have been further encouraged by the fact that the transcendental naturalists, such as John Goodsir and Samuel Brown, who were part of Forbes's Brotherhood, were all influenced by Kantian philosophy through reading the philosophical works of Samuel Taylor Coleridge (Rehbock 1983, 25, 89–91). Also, William Whewell, an influence on Wilson, was one of the leading figures to introduce German idealism into Cambridge. As noted by Robert Bud, Coleridge was the first person to introduce "applied science," an English version of the German term *angewandte Wissenschaft*, in his 1817 *Treatise on Method* (Bud 2012, 538). Kant in his *Prolegomena and Metaphysical Foundations of Natural Science* distinguished between pure science that was based on *a priori* principles and a more practical science that was based on *a posteriori* knowledge gained through experience (Bud 2012, 538–539). This idea of applied science was popularized in Britain through Coleridge's *Encyclopaedia Metropolitana*. In one of the articles, Charles Babbage noted: "The applied sciences, derive their facts from experiments: but the reasoning, on which their chief utility depends, are the province of what is called abstract science" (Bud 2012, 541). This seems close to Wilson's idea that the object of technology is not the practice of art but the principles that underlie art.

Bud notes that about the same time that the term applied science began being used, the "English phrase 'science applied to the arts,' was imported through translation from the French '*science appliquée aux arts*,' the ideological origin of which was quite different from that of the apparently similar 'applied science'" (Bud 2012, 542). The term had its roots in the redefinition of the earlier Conservatoire National des Arts et Métiers (CNAM) which took place under Charles Dupin after he returned to France from Britain where he became influenced by Andrew Ure's idea of "science applied to the industrial arts" (Bud 2012, 542). Dupin was unhappy that the École Polytechnique had moved away from the more practical approach based on the descriptive geometry of Gaspard Monge to the more theoretical and mathematical approach of Pierre Simon Laplace. For Dupin, the CNAM would be a school of the knowledge of science applied to the industrial arts ("*haute école des connaissances des science appliquée aux arts industrielles*") (Bud 2012, 542). The original German term for applied science did not carry the idea of utilitarianism but was used to simply denote scientific knowledge that was gained empirically. On the other hand, the CNAM was designed to have close ties with industry. It contained workshops, a drafting office, laboratories, and a collection of model machines (Alder, 1997, 315–317, 340). Often projects involved designing actual industrial products, such as lock mechanisms for muskets. Therefore, the term science applied to the industrial arts carried with it the idea of industrial applications.

Bud argues that during the 1850s, the distinctions between the German-inspired term, applied science, and the French-inspired term, science applied to the arts, began to dissolve and the terms were often used interchangeably (Bud 2012, 343–345). This may have been partly the result of the Great Exhibition of 1851, which had as one of its main goals, the development of British industry and science was seen as playing an important role. The head of the French delegation to the Great Exhibition was Dupin, who had reformed the CNAM, and Bud notes that during his four-month stay in London, he met at least a dozen times with Playfair. Therefore, when Playfair came to create the Industrial Museum of Scotland and later play a role in creating Wilson's Regius Professorship of Technology, he may have been influenced by the model of the CNAM and its idea of science applied to the industrial arts.

Given the fact that the word technology was almost completely unknown in Great Britain and reflecting the various uses of the word in Germany, France, and America during the nineteenth century that are discussed by Schatzberg (Schatzberg 2018, chap. 6), it is not surprising that Wilson defined technology in a number of ways. Wilson makes two qualifications to his original definition of technology: one that technology includes only the utilitarian arts, and second it includes only certain of those arts. The most important of the arts that was s excluded was medicine because it was so useful it required its own profession (G. Wilson 1855, 4–5, 16). Probably, his earliest definition of technology was contained in a letter he wrote to his sister Jean seemingly responding to her inability to find the word technology in her dictionary. Again, after talking about the Greek roots of the word he says: "*Science in its application to the Useful Arts* is the meaning of the word" (J. Wilson 1866, 279). The fact that technology is defined in terms of useful arts would lead Wilson to sometimes simplify his definition of technology as "the science of Utilitarian Art" (G. Wilson 1855, 16). In a lecture in February of 1856, Wilson defined technology in a number of ways, including the "the science or doctrine of the peculiar practices of the arts," or simply the "the science of art," or "the science of useful art or the useful arts," or "industrial science, *i.e.* the science, or fundamental laws, or guiding principles, or settled theory of the industrial arts" (G. Wilson 1856b, 3). In these various definitions of technology, Wilson seems to be drawing on both French and German ideas. The definition he sent to his sister is very close to the French idea of science applied to the industrial arts, as is his definition of technology in terms of utilitarian art. But Wilson's definition of technology in terms of the fundamental laws, guiding principles, or settled theory seems to draw from the German idea of applied science. In his *Inaugural Lecture*, Wilson expressed disapproval of the term applied science in that it is simply a different way to label art or practice, but as Bud points out, Wilson, in his private correspondence, sometimes did use the term applied science (Bud 2012, 543; G. Wilson 1855, 24–25).

The one common element in all of these definitions of technology is the use of the term science. This does not seem to be accidental. One of the purposes of Wilson's *Inaugural Lecture* was to justify to the faculty and students that this new subject of technology was a suitable subject to be studied in the University (Schatzberg 2018, 92). Sidney Ross has noted that during the first half of the nineteenth century, the word "science" was taking on new meaning and authority (Ross 1962, 65–85). At the end of the eighteenth century and the beginning of the nineteenth century, the words "science" and "philosophy" were often used interchangeably or even in opposite ways that they would be used today. What we call science, especially physics, was labeled natural philosophy, and what we think of as philosophy, especially morals and ethics, were often referred to as the "moral sciences." But, by 1850, the terminology had reversed. Philosophy became associated with the theological and the metaphysical, and science came to refer to the experimental and the physical. Ross notes that the great writer Thomas Carlyle, who had a reputation for identifying the "signs of the times," wrote an anonymous article for the *Edinburgh Review* in 1829, in which he took note of the new explanatory power of the physical sciences. He said:

> It is admitted, on all sides, that the Metaphysical and Moral Sciences are falling into decay, while the Physical are engrossing, every day, more respect and attention... this condition of the two great departments of knowledge; the outer cultivated exclusively on mechanical principles – the inward finally abandoned, because, cultivated on such principles, it is found to yield no result – sufficiently indicates the intellectual bias of our time, ...that except the external, there are no true sciences ...
>
> (Ross 1962, 69–70)

The word science, in its new meaning, was showing up in a number of important works, including Charles Babbage's *Reflections on the Decline of Science in England* (1830), Mary Somerville's *On the Connexion of the Physical Sciences* (1834), and William Whewell's *The Philosophy of the Inductive Sciences* (1840). Another important use of the term science was the creation in 1831 of the British Association for the Advancement of Science (BAAS). It should also be noted that William Whewell introduced the word "scientist" into the English language, first in a review of Somerville's book and then again in his own book (Ross 1962, 71–73). Wilson by prominently using the word science in his definition of technology was clearly drawing on the new power and popularity of the word to justify the study of technology as a worthy university subject. It should be noted that Wilson's use of the term science to define technology could also be seen as reflecting mid-Victorian values of colonialism and empire. If science, in its new meaning, is a defining characteristic of technology then only Western cultures can be thought to have technology. The skills, techniques, and tools that have arisen in cultures without what could be identified as Western science would not seem to count as technologies. I am not sure that Wilson saw things this way since as we will see in a later chapter, he thought it very important that his Industrial Museum collect and display items from Asia, Africa, and the Americas, but his audiences and readers might have assumed a certain Western bias in the definition of technology. Wilson also tried to justify the study of technology as suitable for the University. Wilson notes that his revised definition of technology in terms of the useful or utilitarian arts does exclude the fine arts, which might lead his audience to question how technology would fit into a typical university curriculum that traditionally valued the fine arts. He goes to a great deal of pains to explain his new definition of technology does not imply that the fine arts are useless. In fact, he argues that the fine arts "minister to the wants of the noblest parts of our nature," and that there is a utility in the fine arts, but that utility is attained *"through or by means of Beauty"* (G. Wilson 1855, 5). For Wilson, what distinguishes the utility of technology from the utility of the fine arts is the utility of technology is *"indispensable"* (G. Wilson 1855, 6). He says that the defining characteristic of the utilitarian arts "is not that they deal with what is beautiful or unbeautiful, but with what is *essential* to man's physical existence" (G. Wilson 1855, 6). As is often the case, Wilson explains the relationship between the two types of art in organic terms. He says:

> The Utilitarian Arts thus bear the same relation to the Fine Arts, as affecting our individual condition, which the root, stem, and leaves of a plant, bear to its flower and fruit. The three first are essential to the existence of the plant, and are more or less active throughout its entire life; the flower (including the fruit) is a rare ornament, appearing only at long intervals, sometimes but once in a hundred years.
>
> (G. Wilson 1855, 6)

Again, as with his earlier ideas of unity, Wilson argues that an organic view of the world can lead us to see that there is no opposition between the utilitarian arts and the fine arts, rather they have a symbiotic relationship. In a lecture he gave in February of 1856 to the Philosophical Institution of Edinburgh, Wilson addresses the issue that "utilitarianism is stealing from us our imagination" and that "it is killing our conscience" (G. Wilson 1856b, 5). Wilson's answer is to argue that the goal of industrialism should be to "relieve the wants of the poor" (G. Wilson 1856b, 7). Given his religious beliefs, Wilson always saw his role, and, therefore, his idea of technology, as a moral endeavor. But the relief of the wants of the poor also connected technology to the fine arts since satisfying the needs of food, water, clothing, and housing would allow for the cultivation of the imagination (G. Wilson 1856b, 6). As with the association of technology with science, the association of the fine arts with the leisure that was provided by technology also could be linked to mid-Victorian values of colonialism and empire. It seems to imply that non-industrial civilizations would not be conducive to developing the fine arts since so much time would be spent on simply basic survival.

The idea that technology deals with the essential and indispensable arts gave Wilson another claim for the importance of studying technology. In his inaugural lecture, Wilson argues that technology has to do with the fact that compared to many other animals, humans are "weaker than the weakest of the beasts" and "a hungry, thirsty, restless, quarrelsome, naked animal" (G. Wilson 1855, 7). Wilson goes on to say:

> But it is also the province of Technology to show, that man, because he is this, and just because he is this, is raised by the industrial conquests which he is compelled to achieve, to a place of power and dignity, separating him by an absolutely immeasurable interval from every other animal.
>
> (G. Wilson 1855, 7)

In the same lecture, he would also say:

> With the intellects of angels, and the bodies of earth-worms, we have the power to conquer, and the need to do it. Half the Industrial Arts are the result of our being born without clothes; and the other half of our being born without tools.
>
> (G. Wilson 1855, 9)

He goes on to list the consequences of needing clothes (G. Wilson 1855, 10). In colder climates, early clothing based on animal skins required tools for hunting and additional skills of the tailor and tanner. These skills were expanded into tent-making, which was just a larger type of clothing, and then those skills were expanded to house-making and shipbuilding. In warmer climates, clothing was needed to block the sun rather than for warmth, and this led to clothing based on plants rather than animals, which, in turn, led to planting and reaping along with weaving, baking brewing, and distilling. The use of wood and leaves then led to other types of shelters and buildings. All of the technological developments that arose from needing clothing also required the creation of tools since, unlike the other animals, humans do not come equipped with their own tools. Wilson noted another important characteristic that distinguished human tools from the tools with which animals are born. Like the animals, there are a wide variety of human tools but "they are inseparably dependent on each other" (G. Wilson 1855, 12). The mason needs a carpenter to make a mallet, the carpenter needs a blacksmith to make a saw, the blacksmith needs a smelter to produce iron, and the smelter needs a miner to extract the iron ore. As we have seen in earlier chapters how Wilson saw the natural world as interconnected, he now also seems to see the technological world as interconnected. Our need for clothing requires tools, and those tools require other tools.

In his *Inaugural Lecture*, Wilson provided a final justification for the existence of a Chair of Technology. This justification was based on human nature. Unlike the instincts of animals and insects, human intelligence is fallible. Wilson says: "a strange aspect of imperfection and incompleteness belongs to our human works ..." (G. Wilson 1855, 13). He notes that many times our observations of nature are imperfect as are our applications of the laws of nature to practice. As an example, he discusses how an imperfect knowledge of magnetism led sailors on iron ships to not realize that their compasses were not pointing them in the right direction. He then says: "The deaths of thousands lie at the door of imperfect science; and therefore, the necessity for Industrial Museums and Chairs like this" (G. Wilson 1855, 15). So, an important role of the chair is to root out errors in knowledge and errors in the application of that knowledge.

The fallibility of science led Wilson to two other points. First, the fallibility of human knowledge manifests itself in different ways because of the different endowments of individuals. Some individuals have more knowledge than others. He says:

> We require, accordingly, perpetually to transfer knowledge from the wise to the unwise; from the more wise to the less wise; and such a Chair as this, with its associated Museum, is what, in commercial language, would be called an *entrepot* or exchange for effecting such transfers.
>
> (G. Wilson 1855, 15–16)

Second, human knowledge is "essentially progressive," so new industrial problems continue to arise. For example, the problem with compasses did not arise until shipbuilding began to make use of iron rather than wood. He notes: "Every new discovery and improvement alters the significance and the industrial importance of all which have preceded it ..." (G. Wilson 1855, 16). Given that human knowledge is progressive, it is continually in flux which adds to the need for the Chair of Technology. In seeing technology as progressive, Wilson was probably drawing from the transcendental natural history espoused by Robert Jameson and Edward Forbes.

Wilson also saw technology as a tool to build and maintain the British Empire. In a presidential address he gave to the Royal Scottish Society of Arts on November 23, 1857, he argues that technology will be essential in holding India after the recent Mutiny (now called the First War of Indian Independence) (G. Wilson 1861a, 43–63). He says:

> in one sense, we can hold India only by the sword; but also that it will not be worth holding if we employ no other weapon. Canals railroads, deepened rivers, safe harbours, steam-engines, electric telegraphs, must be spread over it if it is to yield us fruit, or be made a happy land for its people.
>
> (G. Wilson 1861a, 60–61)

But Wilson also engages in some of the typical racist rhetoric of the Empire referring to the "dark and treacherous hands of India," and going on to say that such technological development "must, for a long season at least, be by ourselves, not by the people" (G. Wilson 1861a, 60–61). He then goes on to say:

> European civilization can be carried out only by Europeans. If we are to civilize India, we must not merely give them steam-engines, and the like, but we must stand by our engines, as our soldiers have stood by their guns. The people of India, as a mass, are in the condition in which the Greeks were in the classic ages. They have fine intellects, an exquisite sense of art, and great metaphysical power. But they take no interest in physical phenomena, and are dead to the importance of industrial science.
>
> (G. Wilson 1861a, 61)

This rather blunt racist rhetoric is somewhat unusual for Wilson, especially since his family was antislavery and, as we shall see later, he along with his brother were monogenesis who believed in a single common origin for all humans. His brother Daniel placed great value on the indigenous cultures of North America. Part of the rhetoric may have been the result of the bloody events that had just taken place in India. During the Crimean War, Wilson often expressed shock over the killing that was taking place and thought the end of the Crimean War would bring about a long period of peace, so he was quite dismayed when war again broke out (G. Wilson 1861a, 59–60).

After defining technology and justifying the chair, Wilson had to address one more issue in his *Inaugural Lecture*. As already mentioned, there was the issue of the jealousy of the other professors, especially in the sciences. As Anderson notes, Playfair, who anticipated a possible competition between Wilson and the faculty, broached the idea of Wilson becoming a faculty member with William Gregory, who was Professor of Chemistry and the most likely to be affected by Wilson's courses (Anderson 1992, 177). Gregory gave it his "warm approval" since he thought it would prevent conflicts that would arise if Wilson's lectures were associated only with the Museum, which was outside the University. Ironically, while Gregory welcomed the new position, the most vocal objection to Wilson's role in the University came from J.D. Forbes, the Professor of Natural Philosophy. Anderson quotes a letter written by Forbes two days before Wilson's *Inaugural Lecture* as saying:

> I am surprised to find [the commission] includes no definition of that very vague term [technology] now introduced (for the first time I believe in Britain) as the title of the chair. The word is not to be found in Johnson's Dictionary, and as far as the etymology goes, it might apparently entitle the Professor to lecture on any subject within the faculty of Arts … This is the proper time … for the Senatus to give expression to their opinion of what subjects really ought to be included under this most vague and indefinite commission.
>
> (Anderson 1992, 177)

Anderson notes that Forbes also complained that Wilson was charging less than other professors for matriculating and class fees.

Wilson made light of some of the controversy. Some years earlier, a friend had promised to make him a cushion if he ever received a chair. When he was appointed to the chair, she sent him a handmade cushion and in return, he sent her a poem that he had written entitled "The Chair of Technology and Its Cushion" (J. Wilson 1866, 282–285). In the poem, he talks about the fact that the chair was so new no one understood what it was and where it came from and that he had great difficulty being able to be seated in the chair. He would later say: "The Chair of Technology is not stuffed with down; a thorn or two stick out of it, and it requires cautious engineering to get into it with comfort to myself and others" (J. Wilson 1866, 282–285).

In his *Inaugural Lecture*, Wilson did seriously address the problem of possible conflicts with other faculty. He says: "In short, I have to face the dilemma: – How shall I faithfully fulfil my commission, as Professor of Technology, and yet faithfully respect the rights of my brother professors?" (G. Wilson 1855, 17). Wilson begins by noting that while his definition of technology seems to be broad enough to overlap with other subject areas, that is also true of the subject matter of almost all of the other chairs. Possibly, as a veiled reference to Forbes' complaint about technology, Wilson notes that if natural philosophy is translated to mean the philosophy of nature, it might "amply cover the whole circle of the sciences, and entitle Professor Forbes to discuss, in his own admirable way, every one of them if he pleased" (G. Wilson 1855, 18). But to try to assure the other chairs that he will not be encroaching on their subject matter, he notes that the Chair of Technology has a specific aim. Wilson points out that he is not simply a Professor but also the Director of the Industrial Museum of Scotland. He says there is an "organic connection" between the Chair and the Museum (G. Wilson 1855, 19). This allows Wilson to define his role as Professor of Technology in such a way as to not interfere with the other branches of science. He says: "My office, as Professor of Technology, is to be interpreter of the significance of that Museum, and expositor of its value to you, the Students of this University" (G. Wilson 1855, 19–20). He goes on to say that the Museum will serve as the "text" of his teaching. That is, rather than teaching the theories and principles of pure science, which deals with the natural world, Wilson will be teaching the doctrines or theories of the industrial arts so that if he deals with the subjects of science, he will do so only as they have an application to the industrial arts.

Wilson's definition of technology may have had a lasting impact on America. There has always been a debate concerning how Massachusetts Institute of Technology got its name since it was the first university in America to use the label Institute of Technology. While there does not seem to be any hard, archival evidence about how the name was chosen, Eric Schatzberg makes a strong circumstantial case that can be connected to George Wilson (Schatzberg 2018, 85–94). William Barton Rogers had previously suggested the creation of a polytechnic school in Boston, but by 1860 he labeled it an "Institute of Technology." Schatzberg argues: "Rogers most likely encountered the term in the work of George Wilson ..." (Schatzberg 2018, 92). He notes that Rogers probably encountered Wilson's lectures on *What is Technology?* and *On the Physical Sciences which Form the Basis of Technology*, in the *Edinburgh New Philosophical Journal* which was part of his reading, and Rogers could have also encountered Wilson's idea of technology when he visited the University of Edinburgh in 1857, although there is no evidence the two met. As late as 1863, after MIT had been founded and after Wilson had already died, Rogers was still interested in the "Technology department at Edinburgh" and wrote his brother Henry asking for more information about it (Schatzberg 2018, 93).

Scientific Basis of Technology

Given that Wilson defined technology as the science of the useful or utilitarian arts, his next step was to describe upon what sciences technology would depend on and how it would depend on those sciences. He did this in his introductory lecture of 1856 which he entitled: *On the Physical Sciences which form the Basis of Technology*. He begins by admitting that literally all of the sciences could play some role in technology but argues for focusing on the sciences that have the most direct impact on technology. He essentially divides these sciences into two categories: the naturalistic and the experimental. He defines the first category as dealing with objects, phenomena, and laws that nature "spontaneously" presents to us, while the second category deals with objects, phenomena, and laws that we discover by actively interfering with nature (G. Wilson 1857b, 65–66). Of course, given his belief in the unity of nature, he notes that the distinction between these two categories is only of degree. Wilson then goes on to make some other distinctions between naturalistic sciences and experimental sciences. One distinction is that the naturalistic sciences are also observational, while the experimental sciences are transformational. A further distinction is that the naturalistic sciences are what Wilson calls registrative, meaning that they record or keep track of data about objects, phenomena, and laws, while the experimental sciences are directive, meaning they make use of the direction of forces to bring about transformations. He argues that these two types of science are in some sense two sides of a coin, and industrialists constantly must go back-and-forth making use of both types of science. He says:

> The industrialist must study one class of the physical sciences, or rather one side of all physical science, to consider what gifts nature offers him or her with liberal hand. He must study another class of these sciences, or rather another side of all physical science, to discover how to turn those gifts to account.
>
> (G. Wilson 1857b, 66)

Beginning with the naturalistic sciences, Wilson notes three great benefits of observational science (G. Wilson 1857b, 67–69). The first is that observational science leads to the discovery of new natural substances, such as minerals, vegetable products, and animal products, that have "useful but latent properties" which the transformational sciences can turn into useful products. For example, geology can discover coal and chemistry can produce various gases from it. Mineralogy can discover iron ores that chemists can transform into steel and mechanics can turn into structural bars. Descriptive botanists can discover new wild currants that physiological botanists can cultivate into grapes and chemists can ferment into wines. Descriptive zoology can discover new caterpillars, which physiological zoologist turn into silkworms which are then dyed by chemists and woven into cloth by mechanics. The second benefit of observational science is in discovering new natural phenomena that can be then recorded, or registered, to be used as practical guides. For example, the observation that the sun produces a changing shadow can lead to time-keeping devices. Being able to not only observe but also record phenomena through the use of sextants, thermometers, barometers, anemometers, and weather vanes leads to practical applications in navigation. The third benefit of observational science is to discover new powers and energies in their natural state that can be transformed by the directive sciences. For example, the observational study of meteorology can reveal wind patterns that can be then harnessed by a windmill. While often the observational sciences and the registrative sciences are interconnected, since registration is simply a way to prolong observations of what would be transient phenomena, in many cases, the instruments they employ differ from one another. The goal of observational science is to find the simplest components of complex wholes. Two of the most important instruments to accomplish this are the telescope and the microscope. On the other hand, the instruments of the registrative sciences are to measure the intensity and quantity of phenomena.

In order to discover some of the distinctions between the naturalistic, observational, and registrative sciences on one hand and the experimental, transformational, and directive sciences on the other hand, Wilson begins by comparing astronomy and chemistry as the most characteristic sciences in each category. Before the twentieth-century development of space science with such things as satellites and rovers, the astronomy of Wilson's day was clearly observational since there was no way to transform or experiment on the planets and stars. But Wilson argues that even pure observational astronomy had practical applications the most obvious being the role played by astronomy in navigation (G. Wilson 1857b, 74). By observing the motions of the stars and planets and recording them, navigational almanacs could be developed which would clearly be of use in trade and commerce. In making this argument, Wilson may have been drawing on a controversy between Francis Jeffrey and Dugald Stewart that took place in the early nineteenth century, although Wilson does not directly refer to this debate (Davie 1961, 176). According to George Davie, the debate centered on the utilitarian value of science, especially whether some distinctions could be made between sciences that were based on experiments and sciences that were based purely on observation. The main focus of this debate was Jeffrey's claim that the mental or moral sciences were not scientific since they did not use the experimental method (Tannoch-Bland 1997, 308; Davie 1961, 176). Jeffrey argued that sciences based on experiments allowed for control over nature because they involved manipulations of the natural world, but sciences, like astronomy, that were based purely on observation and did not involve any manipulation of natural objects did not allow for humans to have any power over those objects. Stewart argued against making distinctions between the usefulness of experimental and observational sciences and argued that sciences, like astronomy, could be useful through their use in navigation or in timekeeping.

Wilson also gave examples of other practical uses of astronomy. A particularly important discovery tied to astronomy was Newton's observations of the orbit of the moon, which when combined with his observation of a falling apple led him to formulate his universal theory of gravitation. The new conception of force that emerged from this theory would serve as the basis for a great deal of the mechanical technology that was developed in the nineteenth century (G. Wilson 1857b, 74–76). Wilson thought that one of the most important observations related to astronomy was the fact that the heat, light, and chemical power of the sun, what he called sun-force, could be stored in plants and then released through technology to power both machines and living things. Decaying plants formed coal that became the fuel for steam engines that were powering most nineteenth-century technology so that they were in effect "sun-engines," and that same coal, in the form of charcoal, could be used to smelt ores into things like iron and steel. Plants, like grass, hay, oats, wheat, and other grains could be used to feed animals, such as horses, oxen, and cattle, which could be used as beasts of burden, and Wilson noted that "horse-power is but another name for sun-power." Again, as we have seen before, by seeing machines and animals as all being powered by the sun, he is expressing his cosmic and unitary worldview. He concludes: "Astronomy thus stands much nearer industrialism, in all its departments, than perhaps any of us fully realize" (G. Wilson 1857b, 76).

If astronomy was the epitome of the naturalistic, observational, and registrative sciences, Wilson saw chemistry as the archetype of the experimental, transformational, and directive sciences. He characterized chemistry as "a speechless priestess of nature, sworn to silence, loving concealment, and the most grudging of givers" (G. Wilson 1857b, 76). While the observation of the colors, densities, crystalline shapes, and melting and boiling points of materials can be important and need to be recorded, Wilson sees this as the work of the physicist. The role of the chemist is to question if a material is a compound or is it composed of simple elements. If it is a compound, what are the components of that compound? If it is a simple element, how many simple elements are there? Can these simple elements of compounds be combined in ways not found in nature? Given the secretive nature of chemistry, which has its roots in medieval alchemy, the chemist "must be likened to one of those grim inquisitors of the middle ages, whom no man willingly answered, and who believed in no man's answer unless he wrung it from him by torture" (G. Wilson 1857b, 77). This view reflects Francis Bacon's view that the experimental sciences must interrogate nature. Also like the medieval Inquisition that assumed everyone was guilty unless proven innocent, Wilson says the chemist must regard "substance as 'suspect' or being something else than it seems."

For Wilson, it is not just experimentation that distinguishes chemistry from the other sciences, since, except for astronomy, all of the other sciences engage in some type of experimentation. What ultimately separates chemistry from the other sciences is "its power to modify or transform matter, and to effect the creation of new bodies" (G. Wilson 1857b, 78). That is, chemistry uses an experimental method to bring about a fundamental transformation of substances. Chemists do this through a combination of analysis and synthesis. Wilson lists four different ways in which chemists can transform materials (G. Wilson 1857b, 80–81). First, the chemist can analyze or decompose a substance into its fundamental elements and then use these elements to make other useful materials. For example, salt can be decomposed into sodium and chlorine, and the sodium can be used to make soap, while the chlorine can be used to make bleach. In other cases, the chemist can decompose a material in stages and create useful substance at each stage. For example, rather than simply breaking sugar down to carbon, hydrogen, and oxygen, the chemist can combine the carbon and the water and produce carbonic acid, oxalic acid, alcohol, lactic acid, and various other mixtures. Second, the chemist can combine substances in order to produce rare natural compounds by artificial means. For example, rather than searching for cinnabar, the chemist can make it by combining mercury and sulfur or vermillion dyes can be artificially produced in England, or instead of having to go to India to get natural saltpeter, it can be chemically produced at home. Third, the chemist can combine analysis and synthesis by removing certain elements from a substance and then substituting another in their place. For example, oxygen can be removed from iron ore and carbon put in its place producing cast iron or steel, or through a number of substitutions, a carbonate can be transformed into chloroform. Finally, the chemist can transform a material without removing and adding any new elements, but by rearranging the constituent particles and, therefore, changing the basic properties of the original substance. The simplest example is the transformation of a material from one phase to another. A volatile liquid can be crystallized into a solid, or sugar can be transformed into wood pulp or cellulose.

The Eye and the Hand

Wilson has shown that the industrial sciences depend upon naturalistic, observational, and registrative sciences, like astronomy, and equally on experimental, transformational, and directive sciences, like chemistry. To represent this dependence on both types of science, Wilson proposed a symbol to represent the industrial sciences and be the crest of the Industrial Museum (Swinney 2016, 166). Since the eye could be seen as the symbol of observational sciences like astronomy and the hand the symbol of transformational sciences, like chemistry, Wilson suggested that the "symbol of industrial science is a hand with an eye in the palm, and the fingers free" (G. Wilson 1857b, 84). In a lecture in 1857, he developed his symbol further. He now suggested that the eye and the hand should be surrounded by

> a circle, to imply that the museum represents the industry of the whole world; within the circle, an equilateral triangle, the respective sides of which shall denote the mineral, vegetable, and animal kingdoms, from which industrial art gathers it materials;

And inside the triangle, the eye and the hand (Swinney 2016, 166). The symbol, which has obvious Masonic references, may have been influenced by Wilson's association as a student with Edward Forbes' Universal Brotherhood of Friends of Truth, which has been discussed in an earlier chapter. The organization and symbols of the Brotherhood had many Masonic overtones, and the badge of the Brotherhood was a triangle with the Greek words, oinos, eros, mathesis, or wine, love, and learning on each side of the triangle (J. Wilson 1860, 226–231). Wilson's friend Edward Forbes was known to sign many of his letters throughout his life with a triangle. Unfortunately, since Wilson did not live to see the completion of the Industrial Museum, the symbol was never placed as a crest on the building, but it does appear on Wilson's tombstone (Swinney 2016, 167–168). The symbol also appears on the title page of his lectures and on some of the specimen labels of objects collected for the Museum (Swinney 2016, 177). Geoffrey Swinney, who has both worked at the National Museum of Scotland and written on its history, discovered a glass goblet in the Museum with Wilson symbol on its base (Swinney 2016, 188–190). The goblet was made by W. Keedy in 1858 when Wilson took his students to the Holyrood Glass Works, but it is likely that the symbol was added to the goblet at some later time.

While astronomy and chemistry represented two extremes of industrial sciences, Wilson discussed other sciences that either stood closer to astronomy or closer to chemistry (G. Wilson 1857b, 84–91). The closest science to astronomy was geology which like astronomy humans had limited abilities to transform or direct. Even though humans had limited ways to have a large-scale transformation on the Earth, geology did allow some types of experimentation. Through digging mines and canals, and building roads, railways and breakwaters geologists gain new information about the Earth. Wilson argues that while geology is to a certain degree experimental and transformational, it is not registrative since it cannot predict earthquakes or volcanic activity, and it is not directive since the forces involved with geological activity cannot be used to transform materials for industrial purposes. He concludes that geology "is of the greatest importance, however, to industrialism, in its purely observational character, as dealing with the globe as a great storehouse of mineral matters of the highest value" (G. Wilson 1857b, 86). In Wilson's scheme, the closest experimental science to chemistry is mechanics, which as a science of force also included the study of heat, light, electricity, and magnetism. Mechanics is clearly directional since its main purpose is to direct forces in order to transform materials. It is also clearly transformational since it can transform force into motion and transform one type of force into another, such as transforming heat into motion, or work, transforming electricity into magnetism, or transforming magnetism into electricity, or transforming electricity into heat, or transforming electricity into light. But also, through machines, mechanics can transform materials, such as transforming cotton or wool into textiles. Finally, mechanics is experimental since the invention of things like the steam engine, the cotton gin, or the spinning jenny required a great deal of experimentation.

Wilson argued that mechanics could provide industrialists with transformational power in three ways (G. Wilson 1857b, 87). First, mechanics could provide motive power to set things, such as railway engines, steamboats, and a vast number of machines, into motion. When this motive power is associated with machines, it has tremendous transformational power, turning raw materials such as wool, cotton, silk, and flax into textiles or wood pulp or rags into paper. Second, the mechanical force can also be used to alter the shapes of bodies, through such skills as stone-cutting, wood-carving, and engraving. Probably more important, the mechanical force used in machine tools, such as lathes, drill presses, milling machines, and steam hammers, can actually be used to make new machines. Third, Wilson says that mechanical force can be used to bring about nonchemical internal molecular changes, by which he seems to mean what we would call today, solid-state physics, since the examples he gives are tempering metals, annealing glass, and baking porcelains.

Technology as Biology

The discussion of geology and mechanics leads Wilson to conclude that the sciences that form the basis of industrialism can be arranged in the form of a crescent. Astronomy and next to it geology and next to that mineralogy are on one horn and represent the naturalistic, observational, and registrative sciences. On the other horn are chemistry, and next to it mechanics and next to it heat, light, electricity, and magnetism representing the experimental, transformational, and directive sciences. But there was something missing. There was one science that would sit in the middle of the crescent and give unity to all of the industrial sciences. He says: "In the center of the crescent stands the remarkable science which we have still to consider, namely, biology" (G. Wilson 1857b, 91). Under biology Wilson includes botany and zoology. He notes that as commonly seen as natural history, they are often thought of as purely observational, but he argues: "Nevertheless, every living plant and animal is for the industrialist a machine or apparatus, possessed of remarkable transforming and transmuting powers, which, to a very considerable extent, may be controlled, directed, and even modified by him" (G. Wilson 1857b, 92). Biology gives industrialists a whole new range of powers. Wilson says:

> the fact that, to the extent an organism can be wielded by us, it enables us to add to the transforming and transmuting powers of mechanical and chemical force, which alone are available in the dead machine, the metamorphosing power of vital force.
>
> (G. Wilson 1857b, 92)

This is a truly revolutionary idea since it totally changes our view of technology. Technology is no longer simply based on mechanical transformations and chemical transmutation but now on vital metamorphosis. As an example of how this new biological view would provide a new perspective on technology, Wilson notes that:

> We do not sufficiently remember that all other machines are the offspring of living machines. A steam-engine is the literal as well as the metaphorical embodiment of so much horse-power. A railway viaduct is the petrifaction of so much animal force. A power-loom, after its last improvement, remains still a hand-loom.
>
> (G. Wilson 1857b, 93)

Along with the motive power that vital animal force provides, Wilson noted that plants and animals can provide other useful industrial functions

(G. Wilson 1857b, 93–97). Plants especially can be useful as collectors and manufacturers. For example, as soon as a plant begins to grow, it begins extracting from the soil, air, water and useful chemicals such as alkali or potash that can be valuable as a fertilizer. Humans could extract potash directly from the soil but it would be a costly and laborious process, but plants can extract and store it in small amounts every day and by the end of a growing season, the plants can be burned and the potash can be efficiently released into the ashes. Plants can also extract valuable materials from the sea. Iodine, which is useful in medicine and by photographers of the time, is abundant in seawater but difficult for humans to extract but easy for sea-weeds to absorb, and then the sea-weed can be burned to obtain the iodine and other useful chemicals. Not only can plants collect valuable materials but the animals, like cattle, that eat the plants can accumulate valuable materials in their bones which can be recovered after the animal dies.

According to Wilson, a much more important role for plants and animals is their value in being able to manufacture useful materials and products. A simple acorn can produce, over time, an oak tree which at the time could be used to build naval ships. Other trees provide a vast variety of woods for different purposes, such as ships' masts, timber for building, and mahogany for furniture. In Wilson's time, plants, like the indigo and madder roots, still provided a significant amount of the dyes that were used in textile manufacturing, and plants were the only source of gutta-percha (rubber) used to insulate the trans–Atlantic telegraph lines. Such plants become the ultimate model of a manufacturing process, since they take raw materials and convert them into finished, useful products. It is not by accident that manufacturing facilities have become known as "plants." Not only are plants capable of manufacturing useful products but Wilson notes that animals are just as capable. As examples, he lists the silkworm that turns mulberry leaves into silk, the bee that turns sugar into wax, the oyster that turns sea-chalk into pearls, the turtle that turns sea-weeds into tortoiseshell, and the whale that turns plankton into oil and whalebone. In addition, birds provide us with quills and feathers, elephants with ivory, sheep with wool, cattle with leather, beavers with hat-felt, and other animals provide us with furs. Wilson finally points out how animals provide us with food, not just their meat but items such as milk, butter, and eggs.

Although industrialists in the nineteenth century could not yet construct machines to copy the means by which plants and animals were able to collect and manufacture useful products, Wilson notes that humans did have the singular power to modify plants and animals, and, thus, modify their ability to collect and manufacture useful products. He points out that agricultural shows display a large number of domesticated plants and animals that have very little resemblance to their wild ancestors. He argues: "We have as truly created such fruits and vegetables as the chemist has created ether or chloroform" (G. Wilson 1857b, 97). Wilson sees this ability to modify living organisms as particularly striking when applied to animals. He says:

"Our dogs, horses, and cattle, we have *made*, as truly as we have made glass, or bronze, or porcelain" (G. Wilson 1857b, 97). That is, specialized breeds of dogs, like pointers, or racehorses, or specialized breeds of cattle, are not natural but are the result of our modifications of nature. Wilson notes that a key element in the modification of animals is "pairing in special ways" and later says that a breeder "can exalt or diminish [physiological characteristics] by due selection of sire and dam" (G. Wilson 1857b, 98–99). With these techniques, the breeder "will contract *to make* you a horse according to the pattern you select, as an engineer will make you a steam-engine" (G. Wilson 1857b, 99). Wilson here seems to be referring to the practice of selective breeding or artificial selection that had become popular in the late eighteenth century because of the work of Robert Bakewell. The idea was based on the observation that there are often variations in animals in terms of their physical characteristics, and through careful selection of a male and female breeding pair, those variations can be passed on to the offspring. Charles Darwin refers to artificial selection in the first chapter of *On the Origin of Species* and uses it to lead to his crucial idea of natural selection. There is some debate whether artificial selection actually played a role in Darwin's development of natural selection, or whether he simply used it to justify his new idea.

Although writing before the publication of Darwin's theory, Wilson does make a connection between modifying animals and the idea of evolution. In discussing how animals can be modified, he says: "We do not generally call this *creation*, because we quickly realise that we are but evolving certain germinal tendencies latent in the plants or animals whose offspring our interference renders so unlike themselves..." (G. Wilson 1857b, 98). But he then goes on to make the interesting argument that industrial development might also be seen as a type of evolving. After speaking of plant and animal modifications in terms of evolving, he continues to say:

> but we do no more when we call into existence glass or ultramarine: for unless the elements of these compounds had inevitably tended to produce them under the conditions which we secure, the securing of these conditions would no more have produced them than the mating, under certain restrictions, of particular vegetable or animal pairs would have given us the grapes of Portugal or the race-horses of England.
>
> (G. Wilson 1857b, 98)

Aside from artificial selection, Wilson also notes that plants can be modified through variations in soil, latitude, climate, methods of planting, and types of fertilizers. Here, he again seems to be noting the important role of what we would call the environment for industrial development. In the end, biology has a special place in the scientific basis of industrialism since it is both naturalistic and experimental, both observational and transformational, and both registrative and directive.

In later lectures and publications, Wilson expanded on his biological view of technology. In a paper read to the BAAS entitled "On the Electric Fishes as the Earliest Electric Machines Employed by Mankind" and published in the *Edinburgh New Philosophical Journal* in 1857, Wilson provided a concrete example of how biology could provide a model for technological devices. He noted that the Greeks and Romans use the shocks from electric eels and electric rays (torpedoes) for medicinal purposes, and they were still being used for that purpose in Africa and India (G. Wilson 1857a, 267–287). He concludes that the oldest form of electricity employed by humans was the living electric fish. Not only was the electric fish the oldest electric machine, but also it provided physicists with some important new insights into the nature of electricity. As we have seen, in his book, *The Life and Works of the Honourable Henry Cavendish*, Wilson provides a detailed description of Cavendish's experiments on torpedoes (G. Wilson 1851, 466–470). In his own paper, Wilson concludes that there is not "the slightest doubt, that inorganic electricity, both as a science and art [meaning technology], is very largely indebted to organic electricity, alike for the explanation of the laws which it obeys, and for the contrivances by which it works" (G. Wilson 1857a, 286).

Wilson was not simply content to make practical justifications for the use of biology in the industrial arts, he would go on to make a philosophical argument why technology needed to be dependent upon biology. As discussed in an earlier chapter, Wilson's lecture entitled "On the Character of God" addresses what appears to be a contradiction between the doctrine of final causes, or teleology, and the doctrine of the unity of organization or morphology (G. Wilson 1856a, 26–75). The problem is that when it comes to animal organs and structures, the teleologists want to argue that each individual organ in an animal was designed for a uniquely specific purpose, while the morphologists argue that the organs in an animal were not destined for a specific use to that animal but represent some overall patterns that are present in a wide range of animals (G. Wilson 1856a, 69–71). For example, to a teleologist, the bones of a hand in a monkey or ape are purposely designed for the function of grasping objects and tree limbs. To the morphologist, the bones of a monkey's hand are simply one example of similar patterns of bones that appear in the fins of whales or the flippers of seals that play no role in grasping objects. But in trying to overcome the apparent opposition between teleology and morphology, Wilson notes that just because an organ such as hand-like bones in a whale or seal seems to be useless to those specific animals does not mean that they do not have some higher utility or might have a utility at some different time and place.

From the point of view of the teleologist, a human hand, a lion's paw, a bird's wing, a seal's flipper, and a whale's fin are all specifically designed for a specific purpose such as grasping objects, attacking prey, flying, or swimming, and from the point of view of purpose, they seem to be unconnected to each other. But Wilson then takes the point of view of what he calls the morphological homologist who will argue that each of these organs is simply a slight modification of each other. A wing is simply a hand with elongated fingers, or a hand is simply a wing with shortened fingers. A flipper is simply a hand with shortened and bound fingers, while a hand is a flipper with longer unbound fingers. This leads Wilson to the conclusion that

> the whole of organs in question may be looked upon as modifications of one ideal archetypal form, and many of the structural peculiarities may be explained by reference to this relation, which cannot be accounted for by reference to the use which its living possessor makes of the organ exhibiting them.
>
> (G. Wilson 1856a, 70)

But Wilson cautions against using the morphologist's insights in order to disprove the idea of final cause.

To connect morphology with teleology, Wilson turns to technology (G. Wilson 1856a, 71). He asks us to consider a stranger who had never seen any objects made out of glass and knows nothing of glass-making. After viewing a variety of objects, such as a window pane, a bowl, a finger-glass, a tumbler, a wine glass, a beaker, a bottle, a funnel, a chemist's retort, a barometer, and a thermometer, the stranger would conclude that each object expressed the wisdom, skill, and power of the glassmaker and that each object was designed and created for a specific purpose, so the final cause is clearly discernible. But if at some later time, the stranger is taken to the glassmaker's house to watch him create these various objects, the stranger might be shocked to learn that all of the objects previously admired all began in a common archetypal form of a hollow ball or egg-shaped vessel with a neck like a Florentine flask. The stranger discovers that a thermometer can be produced by elongating the neck of the flask, a barometer is the result of elongating only the neck and suppressing the body of the flask, a funnel is the result of opening the base of the flask, and a bottle, a tumbler, and a bowl are simply elongated flasks with flattened or arched bases. Most surprising would be that a window pane is simply a flask that has been spun around until it becomes an open sheet. Wilson concludes that just as a botanical morphologist can explain different organs in plants as variations of a leaf and a stem, the anatomical morphologist can explain a variety of boney structures in terms of a rib and a vertebra, what he calls the "vitreous morphologist" can trace all objects back to a tube and a ball. Wilson argues that while a wide variety of glass items might all be variations on some simple forms, this leaves open the power and wisdom of the glassblower to use those simple forms to purposefully create a wide range of useful items. Wilson concludes that teleology and morphology can be seen as compatible. The special endowments of single creatures can be seen as "an infinitesimal part of one vast harmonious whole," and yet each separate organism can be seen "as perfectly endowed to fulfill the end of its being ..." (G. Wilson 1856a, 74). If Wilson's main argument is to find a unity between morphology and teleology, another unsaid conclusion is that there is a unity between the biological world and the technological world. That is, glass-making, like the biological world, is based on some archetypal plan.

In one of his last published articles, Wilson would make explicit that technology should be based on biology because the natural world can provide industrialists with divine plans, patterns, and designs (G. Wilson 1859, 279–281). He begins by noting the difference between animals and humans when it comes to creating some artifacts. Although both avoid chance and instead follow some type of prearranged plan, humans attempt to realize an idea that has arisen in their own minds through conscious thought. On the other hand, he argues that animals work by instinct and for Wilson, this animal instinct is an important characteristic. He says:

> the animal artist carries out a conception which is not its own, but a divinely-implanted instinct – in other words, a thought of God's. Each animal instinct is thus equivalent to an infallible recipe or formula of guidance, furnished by God to the creature that follows it; ….
> (G. Wilson 1859, 279)

As such, Wilson argues that human workmen might be able to find better examples of patterns, workmanship, and tools by turning to the biological world for inspiration. And for Wilson, we should not simply turn our attention to animal designs but also to plants that he says have more marvelous transforming powers than do animals. He says: "If, for example, the animal is more wonderful as a mechanician, the plant is more wonderful as a chemist" (G. Wilson 1859, 279). Wilson argues that human industrialists can choose between two modes of design and invention: first, they may rely on the human capacity for creative invention and produce novel devices, or human industrialists may accept the divine patterns found in biology and imitate the workmanship of plants and animals. Wilson is quick to note that human industrialists have always been both inventing and copying.

Wilson then turns to provide some concrete examples of how patterns in nature can lead to new practical or commercial objects. He does note that one has to be careful in assuming that a humanly made object had as its source some natural pattern, since certain forms are so common that similarity does not always imply imitation, but he goes on to argue that the names that are used for certain industrial instruments or vessels can provide some evidence of a connection to natural materials or forms. As examples, he notes that drinking vessels are often called a horn of ale, even though no longer made of an actual horn, or that a brass hunting horn conceals the shape of an actual horn. He notes that the ancient flute-like musical instrument, the *aulos*, is called a *tibia* in Latin and may have been originally made of a leg-bone. Similarly, Vergil's use of the term *avena* for a pastoral pipe may reflect it originated as a reed. He offers that evidence that a modern object is rooted in a pattern of nature could be found by studying other civilizations to see how they produce similar types of vessels.

Wilson then states that modern tubular and hollow vessels have been modeled on three distinct types of natural objects: "Firstly, the stems, leaves, flowers, and fruits of plants; secondly the bones, including the hollow horns of the mammalia; thirdly, the shells or external skeletons of the mollusca" (G. Wilson 1859, 281). He then focuses on the first type of natural objects and shows how gourds and calabashes serve as patterns for useful objects. Gourds and calabashes seem to have provided forms for useful objects for four reasons: first, they are widely distributed over the globe; second, their fruits appear in a great diversity of forms; third, they can be easily made hollow by removing their pulp; fourth, they can be somewhat molded while they are still growing. As examples of the great variety of vessels that can be made based on gourds and calabashes, Wilson points to a number of objects in the Industrial Museum that have been collected from all parts of the world. These included cups, bowls, water-jugs, cooking pots, spoons, scoops, funnels, and cupping instruments. He goes on to conclude that the gourd is the prototype of three hollow vessels. First is a long-necked bottle with an egg-shaped body that has been imitated in clay in China, India, Egypt, and Africa. These types of bottles are so widely known that botanists have used the term "bottle gourd" to refer to the plant. It was but a simple step to imitate the shape of this gourd in ceramics and glass. Second, the hourglass gourd has provided a model for the pilgrim's bottle which can be carried over the shoulder and can be seen in the illustrated version of *Pilgrim's Progress*. Again, the shape of this gourd has been copied in clay and ceramics. Third, gourds have become prototypes of cupping-glasses, which are heated and placed on the skin, often the back, and when they cool, they produce a vacuum that is thought to have some therapeutic benefit. Wilson notes in passing that cupping instruments may be some of the oldest devices that produced artificial vacuums, predating the air-pump.

It seems likely that Wilson's ideas about the relationship between biology and technology were shaped by the concepts of transcendental natural history that he learned from Robert Jameson and then were reinforced through his friendships with Edward Forbes and John Goodsir who both followed that tradition. As noted earlier, transcendental or philosophical natural history drew on Platonism, and German philosophy involved "an a priori belief in the existence of ideal, or 'transcendent,' patterns in nature" (Rehbock 1983, 4). Wilson seems to be following this approach when he claimed that a study of plants and animals could provide industrialists with "God-given" designs or patterns that could serve as the foundation for technological development.

Probably, another influence on Wilson's biological view of technology was natural theology. We have already discussed the widespread influence of natural theology on scientists during the first half of the nineteenth century, but it seems that Wilson made special use of it in arguing that biology could provide patterns or designs for industrialists. He was clearly drawn to natural theology, but as we have seen, he often transformed natural theology into the theology of nature. We have already discussed that in 1848, he published a paper on "Chemistry and Natural Theology," and he referred to William Paley in many of his published works. But, in the paper, discussed above, in which he argued for a connection between the archetypal form that exists in nature and the archetypal forms that serve as industrial designs, he made an interesting use of Paley's design argument. As typically understood, Paley was arguing that the rational patterns and designs that we find in nature would lead us to conclude the existence of a rational God. His classic example is the discovery of a watch would inevitably lead to the conclusion that it was created by some watchmaker and did not just appear spontaneously. This is often referred to as the argument *from* design since the argument is beginning with rational design, a clock, and then concluding with the existence of God, the rational designer. But in his paper on archetypes in nature and technology, Wilson makes a subtle change to the design argument and refers to the argument *for* design (G. Wilson 1856a, 26, 34, 63). While the difference is subtle, it may be particularly important in understanding the biological basis of technology. Wilson was always known for his writing and his careful use of words, so this might not just be a slip, since he uses the phrase several times in the paper. An argument *for* design seems to reverse the usual argument in that instead of beginning with design and then concluding with the existence of a rational God, the argument now seems to be that an a priori belief in the existence of a rational God implies the existence of rational designs in nature and provides a rationale for industrialists turning to biology for inspiration. This could also be seen to reflect a theology of nature more than natural theology. It could also be noted that this reinterpretation of the design argument would also provide a justification for the methodological program of transcendental natural history. As we have seen, Edward Forbes was often critical of the natural theology expressed in the *Bridgewater Treatises* (G. Wilson and Geikie 1861b, 547).

Finally, Wilson used his idea of the connection between technology and biology to organize his set of lectures on technology as part of his duties as both Regius Professor of Technology and Director of the Industrial Museum of Scotland. The lectures extended over three sessions and included an odd mix of topics that would not be seen as typical in a modern engineering curriculum. First, the lectures were organized more around the natural sciences than the physical sciences. According to the University Calendar for 1858–1859, Wilson's course on technology was divided into three sessions covering Mineral Technology, Vegetable Technology, and Animal Technology (Birse 1983, 67–68). Although the division of animal, vegetable, and mineral goes back at least to Aristotle, it was made famous as the basis for Carl Linnaeus's natural system of classification using a binomial taxonomy. His extension of this system from plants to animals and then to minerals gained support from Abraham Werner in his work in geology. We have already seen that Wilson and his friends in the so-called Brotherhood were strongly influenced by Werner's ideas when they were students. The session on Mineral Technology began with what we might call a study of ecology, but what Wilson labeled "Industrial Relations of the Atmosphere, the Waters, and the Solid Crust of the Globe," and then continued with topics on building stones, mortars, coal, and other solid fuels, turning finally to glass-making, pottery, porcelain, and metallurgy (Birse 1983, 67–68). The session on Vegetable Technology began with a discussion of the "Plant as a manufacturing Agent" and then continued with topics on fermentation, distillation, gas-making, the mechanical and chemical applications of wood, candles, soaps, textile materials, bleaching, dyeing, and Calico-printing. The final session on Animal Technology began with a study of the mechanical and chemical applications of bones, horns, shells, and corals and continued with discussions of skins, tanning leather, furs, and wool. As can be seen, the courses were heavily aimed at the role of technology in commercial applications. This focus was continued in Wilson's plans and organization for the Industrial Museum of Scotland, which will be the focus of the next chapter.

The Eye and the Hand, and the Heart

In his final paper, a short popular work entitled "Paper, Pen, and Ink: An Excursus in Technology" that was published in *Macmillan's Magazine* in 1859, Wilson extended his concept of the role of the hand and the eye to also include the heart, a term Wilson used to refer to the emotions (G. Wilson 1859–60, 31–39). In this charming little essay, Wilson makes a study of the tools of writing and their relationship to one another. It could be seen as an example of what Frank Turner has referred to as the *"Baconianism of everyday life"* (Turner 1993, 124). While Turner saw the new focus on the practical aspects of everyday life as contributing to the separation of Bacon's idea of a "double revelation of divine knowledge through both nature and the scriptures," Wilson saw a study of everyday writing tools and providing new insights into the nature of God (Turner 1993, 124). For Wilson, paper, pen, and ink were seen in the broadest terms (G. Wilson 1859–60, 34–39). Paper included all materials on which messages can be stored, even the stars and planets of the universe. A pen was any item that could impress a mark on any material, and ink was every tint or color that could make something visible. He does note that in some cases, such as the blind, messages could be read without ink, and in some cases, two of the items could be combined, such as charcoal functioning as both pen and ink, or a photograph functioning as both paper and ink. But, in general, all three interact together and all three are equal in creating a message. While all three were equal, Wilson saw them functioning in different ways and interpreted each of them in biological terms. Whatever material was used on which to write, the ultimate message was conveyed through the retina, so paper was associated with the eye. Whatever tool was used as a pen was always controlled by the hand. Like his earlier use of the role of the hand and the eye in the sciences that formed the basis of technology, the paper was passive and played a role similar to the naturalistic, observational, and registrative sciences, and the pen was active and played a role similar to the experimental, transformational, and directive sciences. While biology held together all of the sciences, what held together pen and paper was ink that played a role similar to the heart since through the ebb and flow of ink the emotions could be expressed. It is possible that Wilson in his last paper in the last year of his life was thinking about technology in evangelical terms. As we saw in an earlier chapter, Thomas Chalmers argued that the natural sciences could provide evidence of the existence of God, but it was only through a study of the mind that one could understand the moral character of God. By combining elements of natural theology with a study of the mind, one could develop natural theology of conscience. It seems that Wilson may be arguing in this paper that the technology of writing cannot be simply understood in terms of pen and paper, hand and eye but requires an understanding of the heart or emotions.

At the very end of the paper, Wilson seems to be referencing some of Chalmers' other evangelical theology. As noted earlier, Chalmers' main criticism of natural theology was that it focused too much on good and not enough on evil or original sin. He often focused on depravity, original sin, and emphasized "principles of destruction." Wilson ends his paper drawing upon newly scientific discoveries to raise the issue of an ultimate principle of destruction. While noting that human writing materials are not eternal, he refers to the new discovery of William Thompson (later Lord Kelvin), Hermann von Helmholtz, and John Tyndall. Although he does not use the term entropy or refer to the second law of thermodynamics, it is clear that this is what he is talking about. He notes that while God's universe appears to be eternal:

> But in our own day and amongst ourselves has arisen a philosopher [William Thompson] to show us, as a result simply of physical forces working as we observe them to do, that the lettered firmament of heaven will one day see all its scattered stars fall like ruined type-setting of a printer into one mingled mass.
>
> (G. Wilson 1859–60, 39)

As a result, Wilson concludes that "It is the writer that shall be immortal, not the writing" (G. Wilson 1859–1860, 390). Therefore, as a result of a study of the everyday tools of writing, Wilson is able to discover that other aspect of Bacon's double revelation – the one that comes from scripture. While much of Wilson's work on the Industrial Museum of Scotland would be more focused on the Baconianism of everyday life, his last paper seems to indicate that he continued to search for Bacon's double revelation of divine knowledge.

References

Alder, Ken. 1997. *Engineering the Revolution: Arms and Enlightenment France, 1763–1815*. Princeton, NJ: Princeton University Press.

Anderson, Robert G.W. 1992. "'What Is Technology?': Education through Museums in the Mid-Nineteenth Century." *The British Journal for the History of Science* 25: 169–184.

Balfour, John. 1860. *Biographical Sketch of the Late George Wilson, M.D.* Edinburgh: Murray and Gibb.

Birse, Ronald M. 1983. *Engineering at Edinburgh University: A Short History, 1637–1983.* Edinburgh: School of Engineering, University of Edinburgh.

Bud, Robert. 2012. "'Applied Science:' A Phrase in Search of a Meaning." *Isis* 103: 537–545.

Channell, David F. 1982. "The Harmony of Theory and Practice: The Engineering Science of W.J.M. Rankine." *Technology and Culture* 23: 39–52.

Constable, Thomas. 1877. *Memoir of Lewis D.B. Gordon, F.R.S.E.* Edinburgh: For Private Circulation.

Davie, George Elder. 1961. *The Democratic Intellect: Scotland and Her Universities in the Nineteenth Century.* Edinburgh: Edinburgh University Press.

1952. *Fortuna Domus: A Series of Lecture Delivered at the University of Glasgow in Commemoration of the Fifth Centenary of Its Foundation.* Glasgow: University of Glasgow Press.

Gladstone, John H. 1860. "Estimate of Literary Character." In *Memoir of George Wilson*, edited by Jessie Aitken Wilson, 509–522. Edinburgh: Edmonston and Douglas.

Rehbock. Philip F. 1983. *The Philosophical Naturalists: Themes in Early Nineteenth-Century British Biology.* Madison: University of Wisconsin Press.

Ross, Sydney. 1962. "Scientists: The Story of a Word." *Annals of Science* 18: 65–85.

Schatzberg, Eric. 2018. *Technology: Critical History of a Concept.* Chicago: University of Chicago Press.

Swinney, Geoffrey N. 2016. "George Wilson's Map of Technology: Giving Shape to the 'Industrial Arts' in Mid-Nineteenth-Century Edinburgh." *Journal of Scottish Historical Studies* 36: 165–190.

Tannoch-Bland, Jennifer. 1997. "Dugald Steward on Intellectual Character." *British Journal of the History of Science* 30: 307–320.

Turner, Frank M. 1993. *Contesting Cultural Authority: Essays in Victorian Intellectual Life.* Cambridge: Cambridge University Press.

Wilson, George. 1851. *The Life and Works of the Honourable Henry Cavendish.* London: The Cavendish Society.

Wilson, George. 1855. *What Is Technology?* Edinburgh: Sutherland and Knox.

Wilson, George. 1856a. "On the Character of God, as Inferred from the Study of Human Anatomy." In *Addresses to Medical Students Delivered at the Instance of the Edinburgh Medical Missionary Society*, 26–75. Edinburgh: Adam and Charles Black.

Wilson, George. 1856b. *On the Objects of Technology and Industrial Museums.* Edinburgh: Sutherland and Knox.

Wilson, George. 1857a. "On the Electric Fishes as the Earliest Electrical Machines Employed by Mankind." *Edinburgh New Philosophical Journal*, new series 6: 267–287.

Wilson, George. 1857b. "On the Physical Sciences Which Form the Basis of Technology." *Edinburgh New Philosophical Review*, new series 5: 64–101.

Wilson, George. 1859. "On the Fruits of the Cucurbitaceae and Crescentiaceae, as the Original Models of Various Clay, Glass, Metallic, and Other Hollow or Tubular Vessels and Instruments Employed in the Arts." *Edinburgh New Philosophical Journal*, new series 10: 279–281.

Wilson, George. 1859–1860. "Paper, Pen, and Ink: An Excursus in Technology." *Macmillan's Magazine* 1: 31–39.

Wilson, George. 1861a. "Address as President of the Royal Scottish Society of Arts at its Annual Meeting, November 23, 1857." *Transactions of the Royal Scottish Society of Arts* 5:43–63.

Wilson, George and Archibald Geikie. 1861b. *Memoir of Edward Forbes, F.R.S.* London: Macmillan and Company.

Wilson, Jessie Aitken, 1860. *Memoir of George Wilson.* Edinburgh: Edmonston and Douglas.

Wilson, Jessie Aitken. 1866. *Memoir of George Wilson*, new edition. London: Macmillan and Co.

Yanni, Carla. 1999. *Nature's Museums: Victorian Science and the Architecture of Display.* Baltimore, MD: Johns Hopkins University Press.

6   Wilson's Industrial Museum

On his birthday in 1855 (February 21st), George Wilson was officially appointed as the founding Director of the Industrial Museum of Scotland (currently the National Museum of Scotland) (J. Wilson 1866, 271). He had heard some rumblings of this back in April of 1854, but nothing seemed to come of it so it came as somewhat of a surprise. At first, he had some reluctance to accept the position because he was still recovering from another health crisis. As we have seen, Wilson suffered from a number of health problems, including partial amputation of his foot, severe rheumatism, and tuberculosis. In early February, he began spitting up blood and continued to hemorrhage for three more days. He assumed that it was connected to his tuberculosis, but it was later found to have come from an enlargement of the spleen (J. Wilson 1866, 269–270). Before accepting the Directorship of the Museum, Wilson consulted with a number of his medical friends who assured him that his health would most likely improve and that it should not interfere with his taking on the new duties.

British Museums

The creation of the new Industrial Museum at Scotland came at a time when the idea and purpose of museums were changing. These changes were the result of a number of factors including the rise of industrialization, political unrest on both the Continent and in Ireland, and the growth of the British empire, and museums were responding and being transformed by all of these forces. Before the nineteenth century, the closest things to museums were the seventeenth and eighteenth centuries' *Wunderkammer*, or cabinets of curiosities (Swinney 2013, chap. 2 n. 4, 5; Yanni 1999, 14–19). These collections included a mix of animal, vegetable, and mineral objects, such as stuffed crocodiles, birds, fish, and other animals along with turtle shells, sea shells, starfish, and horns and antlers and could also include bows, arrows, paddles, and footwear (Yanni 1999, 17–18). The curiosity came from the many objects that seemed to be unnatural, such as a two-headed cat. Such objects raised questions about both the perfection of the natural world and God's control over the world (Yanni 1999, 16). These collections were usually private and to some degree expressed the power and

DOI: 10.4324/9781003212218-6

prestige of the owners. Scholars have debated if cabinets of curiosities are actual precursors of natural history museums. Douglas Crimp has argued that the lack of both focused collecting and lack of any organizational or classification system distinguished them from modern museums, but Paula Findlen responds that the early collectors of curiosities helped to "popularize the study of nature – 'science' in the broadest sense – for the urban elite through their willingness to make learning a form of display" (Yanni 1999, 20).

Some of these collections of curiosities did form the basis of what would later be thought of as something close to the nineteenth-century idea of a museum. In the middle of the seventeenth century, John Tradescant accumulated a large collection of natural objects, including whales' teeth, shells, birds from India, and piece of the supposedly "true cross" (Yanni 1999, 20–24). By later in the century, Elias Ashmole managed to acquire the collection under some questionable circumstances and left the collection to the University of Oxford which opened the Ashmolean Museum in 1683 which was open to all for a small entrance fee. But like the earlier cabinets of curiosities, the Ashmolean Museum was a mix of works of nature, works of humans, and works of God. Also, unlike later museums, the cabinets of curiosities often mixed together natural objects and cultural objects. For example, museums might contain a shell that had been carved into a drinking vessel, or a piece of mineral that had a religious scene carved into it.

A movement toward a more public museum came when Hans Sloane donated his vast collection of objects, including a large number of stuffed animals, to the British people which would serve as the basis for the British Museum (Yanni 1999, 24). It was not until the 1820s that a building would be constructed in Bloomsbury as the permanent home for the Museum. From its opening to the public in 1753 until the permanent building was completed, Sloane's collection was housed in an already existing building. Although technically open to the public, historian Richard Altick notes that it was often difficult to get tickets and the Museum was often closed at times when workers might be free to visit (Yanni 1999, 24). There were other museums dedicated to natural history that opened in the first half of the nineteenth century such as William Bullock's Egyptian Hall which opened in 1812 in Piccadilly and the Hunterian Museum which opened in 1813 in Lincoln's Inn Fields. But Bullock's museum was primarily aimed at entertainment and profit, and the Hunterian Museum was part of the Royal College of Surgeons and primarily open to physicians.

The Crystal Palace

The major impetus behind a changing view of museums and a stimulus to building a number of new museums in Britain was the Great Exhibition of the Works of Industry of All Nations which opened in 1851 in the Crystal Palace that had been constructed in Hyde Park. The exhibition had been conceived of by Henry Cole and Prince Albert. Cole had had a number of governmental positions including work in the Public Record Office and played a role in introducing the "Penny Post," the first postage stamps and the first Christmas cards. But he was also interested in design and had won prizes from the Royal Society for the Encouragement of Arts, Manufactures and Commerce (Royal Society of Arts) which had Prince Albert as a patron (Kriegel 2007, 87). Working together with the backing of the Royal Society of Arts Cole and Prince Albert put forward the idea of a grand exhibition of the products of arts and industry from across the globe. There had been some tradition, especially in France, of national exhibitions of manufacturers but nothing on the scale conceived by Cole and Prince Albert. This would be the first in a series of World's Fairs.

The Exhibition, running from May 1, 1851 to October 15, 1851 attracted six million visitors, including Lewis Carroll, Charles Darwin, Charles Dickens, George Eliot, Karl Marx, Alfred Tennyson, and William Thackery. There were 14,000 exhibitors providing 100,000 exhibits, about half of which came from Great Britain or the British Empire. Famous exhibits included Samuel Colt's repeating revolver, Cyrus McCormick's Virginia Reaper, and Charles Goodyear's Indian rubber. The entire exhibition was self-financing and ended up making a small profit. A big attraction was the machinery that was powering the Industrial Revolution, including steam engines and textile machines. Because of this, there is a common perception that the Great Exhibition represented a triumph of technology, but Lara Kriegel has made the argument that many of the exhibitions represented traditional handcrafts and artisanal labor (Kriegel 2007, 9–10).

The Crystal Palace Exhibition served a variety of purposes. First it aimed to indicate to the world that Great Britain was the world's industrial leader. This seemed under challenge after a successful French Industrial Exhibition in 1844. Prince Albert also believed that bringing together new inventions from around the globe would stimulate the development of technology, and this development would help to establish world peace. Ironically almost two years to the day after the Great Exhibition closed, the Crimean War began. For Prince Albert, exhibitions that focused on the natural world or on industry also had a moral purpose. Shortly before the actual opening of the Great Exhibition, Prince Albert gave a speech in which he said:

> We are living at a period of most wonderful transition, which tends rapidly to the accomplishment of that great end to which, indeed, all history points – the realization of the unity of mankind…. His reason being created after the image of God, he [mankind] has to use it to discover the laws by which the Almighty governs his creation and, by making these laws his standard of action, to conquer nature to his use – himself a divine instrument.
>
> (Swinney 2013, chap. 2, n. 17)

Again, ironically, a number of scholars today would point to the Great Exhibition as both reflecting and furthering the goals of colonialism and empire.

Department of Science and Art

Cole had less grandiose goals for the Great Exhibition. Kriegel argues that as someone engaged in design, he saw it as an opportunity to push forward a plan for the reform of British design. A parliamentary report from the Select Committee on Arts and Manufacturers in 1835–1836 had raised concerns about the work of artisanal designers in Great Britain. Kriegel notes that those who testified had "complained of shoddy calicoes, inelegant cutlery and garish wallpaper" (Kriegel 2007, 2). While British manufactured products could compete with other countries, British designs were seen as inferior. The solution to this problem was thought to be the creation, in 1837, of the government sponsored School of Design which Henry Cole took over in 1849, but the school did not lead to the design reforms that Cole had intended. While the Great Exhibition demonstrated Britain's industrial superiority, it also showed that when it came to aesthetics Britain was still behind. In order to address this continuing problem, Cole came to the belief that design could be improved by improving consumers' taste so that they would demand better designed objects. He believed this could best be done by creating a museum of ornamental art using many of the objects that were left over from the Great Exhibition and the idea gained the support of Prince Albert (Kriegel 2007, chap. 4). The new museum would open in its permanent home in 1857 under the name the name the South Kensington Museum and would become the forerunner of the Victoria and Albert Museum.

The South Kensington Museum would become simply one piece in a larger project of museum building under Cole. The Great Exhibition had been very successful on a number of fronts but it was temporary. The original Crystal Palace was rebuilt in the south London suburb of Sydenham in a different style from the original. It did contain a number of exhibits, including stuffed animals and geological exhibits but it also contained a large organ for concerts and at one point hosted a circus, so it really did not function as a true museum. It eventually burned to the ground in a spectacular fire in 1936. Given the success of the original Great Exhibition Cole began to think of creating other museums around Great Britain. In 1852, the same year the Museum of Ornamental Art opened in temporary quarters, the government created as a sub-division of the Board of Trade, the Department of Practical Arts which was renamed the Department of Science and Art the next year. The Department was partially funded by the profits from the Great Exhibition and was headquartered in South Kensington. Its purpose was to give support to education in science and art along with the practical arts of technology and design. Henry Cole became the Superintendent of the Department and the Scottish chemist Lyon Playfair served as Secretary of Science.

An institution that would come under the control of the Department of Science and Art was the Museum of Practical Geology which would play a role in the organization of the Industrial Museum of Scotland (Yanni 1999, 51–58). Established in 1835 and opened in 1837 as the Museum of Economic

Geology and later renamed the Museum of Practical Geology, it served as a branch of the newly created Geological Survey. The first nation-wide geological survey, its purpose was to map the geology of Great Britain in order to identify and locate useful minerals that might be of value for industry. The Survey was headed by Henry de la Beche and he suggested establishing a museum to assist with the Survey. The purpose of the Museum was to catalog and display objects collected by the Geological Survey as well as maps generated by the Survey. In 1853, the Geological Survey and with it the Museum of Practical Geology was transferred to the Department of Science and Art. Obviously given both titles the Museum was intended to focus on the economic and practical applications of British minerals. Along with the Geological Survey, the Museum was also connected to the School of Mines and the Mining Record Office (which were also transferred to the Department of Science and Art in 1853), so it combined a number of elements beyond the traditional museum and would influence Wilson's idea for the Industrial Museum.

The Museum began in small quarters in Craig's Court, but it quickly outgrew that location and a new larger building was constructed and opened in 1851 on the corner of Piccadilly and Jermyn Streets and came to be known as the Jermyn Street Museum. The Museum represented a new idea for the role of museums in British society. At its opening Prince Albert stated:

> it is impossible to estimate too highly the advantages to be derived from an institution like this, intended to direct the researches of science, and to apply their results to the development of the immense mineral riches granted by the bounty of Providence to our isles, and their numerous colonial dependencies.
>
> (Yanni 1999, 52)

Prince Albert also saw the Museum as playing a role in showing how God had favored Great Britain by depositing coal where is could be most usefully applied to industry and would demonstrate promise of colonialism by displaying objects from across the empire. Another important role of the new Jermyn Street Museum was education. Much of the ground floor was taken up by a lecture hall. During the day, the lecture hall was used by the School of Mines, but at night, it served as a venue for evening lectures aimed at workers and mechanics. Carla Yanni notes that in 1853, the Evening Lectures for Working Men filled the 600 allotted seats in just two days, and in 1852, Lyon Playfair filled the lecture hall, so when Wilson was planning the Industrial Museum of Scotland, he copied such a system of lectures (Yanni 1999, 56–57).

A close connection would be established between the Museum of Practical Geology and George Wilson when in 1844, Edward Forbes joined the Geological Survey and took over the Museum. De la Beche thought having a paleontologist might be useful for the Survey since the identification and study of fossils could provide information about the geology of a region so Forbes was hired as the paleontologist for the Survey and placed in charge of the Museum. As we have seen, Wilson and Forbes had been friends since medical school and kept in close contact and met almost every year at the "Red Lions" dinners at the BAAS meetings (G. Wilson and Geikie 1861, 378). This bond would become even more important when after the death of Robert Jameson in 1854, Forbes was named Regius Professor of Natural History at the University of Edinburgh and with it the Keeper of the Natural History Museum. Unfortunately, Forbes' unexpected death a few months after taking the chair would prevent him from playing much of a direct role in the creation of the Industrial Museum of Scotland. Even though Forbes died before Wilson received his appointment as Director of the Industrial Museum, there is reason to believe that the two might have talked about museums since Forbes had experience heading the Museum of Practical Geology and Forbes arrived in Edinburgh during the summer of 1854. We have already noted that Wilson had some discussions about the Industrial Museum as early as April of 1854, so it is hard to believe that they did not have some discussions about museums before Forbes died in November.

The Educational Use of Museums

Even if the two did not talk directly about museums, Forbes had published an article "On the Educational Uses of Museums," which was the opening lecture to the School of Mines for the 1853–1854 session and Wilson could have easily obtained a copy of the lecture (Forbes, 1853). In his lecture, Forbes makes a strong case for museums as a moral force in British society, which is one of the points we have seen made by Prince Albert. But for Forbes, the moral force of museums was not simply to show the "unity of mankind" or the benevolence that God had placed on Great Britain, rather he believed that museums could change human values. This was particularly important at mid-century when the French Revolution was not that far in the past, other revolutions were continuing on the Continent, and there was serious concern about possible rebellions in Ireland. Michel Foucault is famous for showing how eighteenth- and nineteenth-century institutions, like the prison and the clinic/hospital, were subtly designed to manage and control democratic populations (Kriegel 2007, 5, 161). Some scholars have seen this to be also true of museums since in *The Order of Things*, Foucault argues how the new emerging nineteenth-century systems of classification had built-in hierarchies (Yanni 1999, 8; Foucault 1973, chap. 5).

In his lecture, Forbes makes a clear connection between the order of museums and change in human behavior. In terms of the working classes, he says:

> The labourer who spends his holiday in a walk through the British Museum, cannot fail to come away with a strong and reverential sense of

the extent of knowledge possessed by his fellow-men. It is not the objects themselves that he sees there and wonders at, that make this impression, so much as the order and evident science which he cannot but recognize in the manner in which they are grouped and arranged. He learns that there is a meaning and a value in every object however magnificent, and that there is a way of looking at things common and rare distinct from the regarding of them as useless, useful, or curious – the three terms of classification in favour with the ignorant.

(Forbes 1853, 9)

Explicitly stating how this will change the laborer, he says that after visiting a museum and while at home and during walks, the laborer will "acquire a new interest in stones, in the flowers, in the creatures of all kind that throng around him" (Forbes 1853, 10). But the museum will do more – it will turn the laborer away from vices that are harmful to society. Forbes says the laborer will have "gained a new sense – a thirst for natural knowledge, one promising to quench the thirst for beer and vicious excitements that tortured him of old" (Forbes 1853, 10). As a result, Forbes claims the laborer will "become a better citizen and a happier man" (Forbes 1853, 10).

Forbes also believed that the museum could change the "educated classes" because the "great defect of our system of education is the neglect of the *educating* of the observing powers – a very distinct matter, be it noted, from scientific and industrial instruction" (Forbes 1853, 10). He argues that it is a common error to believe that taste and reasoning that are acquired through literary studies and logic and mathematics are opposed to training in observation through the sciences that are descriptive and observational. Instead, he claims that both are needed – literature, mathematics, and logic to understand the forms and observation to understand the phenomena, and this can be done through education in practical applications and that "museums are the best text-books" (Forbes 1853, 10–11). He concludes that to accomplish this, "the value of Museums must in great measure depend on the perfection of their arrangement and the leading ideas regulating the classification of their contents," so that an educated youth ought to be able to "instruct himself" (Forbes 1853, 11).

Scottish Museums

All of this provides a context to analyze the creation of the Industrial Museum of Scotland. While the Great Exhibition and the subsequent establishment of the Department of Science and Art are the most direct impetus for the creation of the Museum, there was a movement to establish a national museum in Edinburgh before the Great Exhibition (Swinney 2013, chap. 3). During the 1840s, Adam White, who was born in Edinburgh but was working was a zoologist for the British Museum, wrote a series of letters, under the nom de plume, Arachnophilus, to a number of newspapers, including *The*

Scotsman, lobbying for a government-funded museum in Edinburgh. Of such a museum, he said:

> A National Museum must not be linked to Natural History; let it be co-extensive with Art and Science – let it be a nucleus to which the spirited sons of Scotia may give and bequeath pictures, statues, specimens, books, and MSS, – let it be a place to which your hard working sailors, soldiers, merchants, and medical men in active foreign service, may delight to send specimens of Natural History, or curiosities connected with rude and less civilized nations – let it contain a large collection of casts from antiques for artists and architects to copy and study, – let it contain models of the geological structure of your country, which, in itself … is almost 'an epitome of the world'.
> (Swinney 2013, chap. 3, n. 12; Anderson 1992, 174)

White's view of a museum in Edinburgh contained a number of elements. First it seemed to have elements of the old cabinets of curiosities, especially since he uses the term curiosities. It also seemed to reflect the idea of the superiority of the British Empire, especially in its reference to the "rude and less civilized nations." Finally, White expressed the argument that museums could also serve as instruments of national pride as the Great Exhibition would for all of Great Britain, but a major museum in Edinburgh could also serve as a symbol of Scottish pride and nationalism. In this vein, Charles Waterston has argued that the desire for a museum in Scotland was spurred by the establishment of the Irish Museum of Economic Geology (later renamed the Museum of Irish Industry) in 1845 which led to the feeling of many Scots that the government in London was favoring Ireland and ignoring Scotland (Swinney 2013, chap. 3, n. 13, 14).

At the time Edinburgh did have a number of museums, including the Museum of the Royal College of Surgeons of Edinburgh, the Museum of the Society of Antiquaries, and the Free Church Museum, and some art galleries, but most of these were not open to the public or had restricted access. There were two major museums connected to natural history – the Natural History Museum of the University and the Museum of the Highland and Agricultural Society, both of which would play roles in the creation of the Industrial Museum. The Highland and Agricultural Society had been amassing a large collection of minerals, rocks, and building stones, but it was causing a financial drain on the Society, and in 1851, they offered to donate their collection to a proposed new museum in Edinburgh. The Directors of the Society argued that the new museum should be modeled on the Museum of Irish Industry in Dublin rather than the University's Natural History Museum. Saying of the Natural History Museum:

> such a general collection, embracing everything that falls within the comprehensive science of Natural History, is different both in character and object from what has been established in Ireland, and not calculated

to serve the practical purposes in view. They [the Directors] also think that the strength claim was in adhering to the case of Ireland, and craving a similar boon.

(Anderson 1992, 175)

Soon after the Royal Scottish Society of Arts wrote to London seconding the request for a Museum of Economic Geology in Edinburgh. The Royal College of Surgeons of Edinburgh argued in favor of a new museum on the economic grounds of gaining knowledge of Scotland's raw materials. They said:

The establishment of such an institution in the capital of Scotland would be a great national benefit by affording the means of obtaining definite information in regards to the mineral wealth of the kingdom, its ores and coals, its building, paving, and ornamental stones, granites, and marbles, the localities and composition of soils, the qualities and capabilities of its different clays for bricks, tiles or pottery wares, and of its limestones for building purposes and manure, and generally as a means of developing the industrial resources of its territorial products.

(Swinney 2013, chap. 3, n. 33)

Not everyone was in favor of a new museum, given that the University already had a Museum of Natural History. In the late seventeenth century, Toun College, which would become the University of Edinburgh, already had a natural history collection, and under Robert Jameson, Professor of Natural History, the collection grew considerably to the point that in 1820, the University built a new three-story building to house the collection (Swinney 2013, chap. 3, n. 21–25). Although there was some talk of making the collection open to the public, Jameson exerted strict controls and entrance fees. Wilson notes that as a boy he enjoyed visiting the Museum but that he often could not afford the admission price (Wilson and Geikie 1861, 110). When the public was freely admitted on the occasion of Queen Victoria's coronation, there were huge crowds. The Museum was also opened to workers on New Year's Day in 1852 in an attempt to limit drinking among the workers (Swinney 2013, chap. 3, n. 30). Not surprisingly Jameson was not a supporter of a new museum in Edinburgh seeing as an attempt to establish a New School of Geology and Practical Chemistry that would threaten his position. In a report to the University Senate, he said:

This project I consider hostile to the University, because there are already in our Establishment Professors of Geology and Practical Chemistry, provided with most extensive Museums of Practical and Theoretical Geology and of Practical Chemistry. I may further add that this project is prejudicial, as it interferes with our more legitimate claims for pecuniary aid for the extension of the Museum buildings.

(Anderson 1992, 175)

Instead of a new museum, Jameson got his colleagues to write to the Lords of the Treasury opposing a new national museum and instead expanding the University Museum to four times its size.

The government rejected Jameson's suggestion, but the University moved forward with a new bold plan. A memorandum noted that Edinburgh could not support a new National Museum of Natural History along with the University Museum and therefore suggest combining the two museums into one "by proposing to contribute all of available resources at their [the University] command for converting the University Museum of Natural History into a National and open one" (Swinney 2013, chap. 3, n. 36). Although giving up its natural history collection, the University wanted to maintain some control and argued that the Regius Professor of Natural History should become the new Museum's Keeper of Natural History, and that courses taught at the new Museum would be part of the University's system of tuition (Swinney 2013, chap. 3, n. 37). These two requirements would establish a connection between the new Museum and the University.

While this proposal would result in the establishment of a new government Museum of Natural History in Edinburgh, it did not address the concerns of those lobbying for some type practical museum similar to the Museum of Practical Geology in London or the Museum of Irish Industry in Dublin. At this point, the government stepped in and placed the issue in the hands of the Department of Science and Art which already controlled the museums in London and Dublin, along with other museums. Lyon Playfair who had become the Secretary of Science in the Department of Science and Art and who although born in Bengal considered himself Scottish and had been educated at the University of Edinburgh became a staunch support of a new Museum for Edinburgh. Playfair's idea for the museum was somewhat unusual and would result in some organizational vagueness. As he saw it the new national Museum would actually consist of two parts: one part would be a natural history museum based on collection of the University that was given over to the government, but there would be a second practical part of the Museum similar to the Museum of Practical Geology in London and the Museum of Irish Industry in Dublin but expanded beyond practical geology to cover other industrial developments (Anderson 1992, 175). This combination of two museums was justified by a statement from the Department of Science and Art which said:

> The fact of the natural history collections forming part of the new Museum will give an advantage to the new institution which few places possess. A scientific museum showing the mode of occurrence of the objects which are afterwards applied in industry, forms a most important step in the efficient study of technical collections. In London, collections of this kind are much dispersed, and can only be studied in their necessary connexions with great inconvenience. … In the new Museum in Edinburgh the scientific and technical collections will be under the

same roof and one management, and may be made materially to support each other.

(Swinney 2013, chap. 5, n. 5)

The physical and organizational relationship between the two parts of the Museum was left to be resolved after the completion of a permanent building to house the Museum (Swinney 2016, 170). But Playfair always saw a close connection between the new Museum and the University which was controlled by the Edinburgh Town Council. He noted:

> Though the Patrons of the University [the Town Council] would have no controlling power in the management of the Museum, the Board of Trade would no doubt think that the Museum was used in the best possible way for the advantage of the public if the Professors of the University employed its collections in the illustrations of the Courses with the consent of the Directors of the Museum.
>
> (Anderson 1992, 175)

The Senate of the University accepted Playfair's proposal with the condition that the Regius Professor of Natural History would become the Keeper of the new Natural History Museum even though the collection would no longer be owned by the University.

The Industrial Museum of Scotland

There was still the issue of what would become the Industrial Museum of Scotland. As with the Natural History Museum, there was some concern among the faculty that the Industrial Museum might infringe on the University faculty. In a letter to Playfair written my Robert Chistison, speaking for the University Senate, he said:

> the Keeper of the Technological department shall be excluded from converting his Lectures on the Technology of Natural History into scientific lectures on Systematic Natural History —and conversely they [the Senate] think it will be advisable, also, to secure as far as possible any injurious interference on the part of the Keeper of the Technological department of the Museum with the proper province of the University Professor of Chemistry.
>
> (Anderson 1992, 176)

As we saw in the last chapter, this was one of the first issues that Wilson addressed in his *Inaugural Lecture*. Also, as we saw the University moved to have some control over the Industrial Museum by creating a Regius Professorship of Technology for Wilson, so George Allman, the then Regius Professor of Natural History at the University, would also serve as Keeper of

Natural History in the Natural History Museum, and Wilson, as Director of the Industrial Museum, would also serve as Regius Professor of Technology in the University. The actual relationship between the two was to be determined once the actual Museum had been built.

A minor problem arose when at the last minute the city of Glasgow argued that it should be the home of the Museum, but when the Duke of Hamilton, the Duke of Buccleuch, the Duke of Montrose, the Earl of Eglington, and the Earl of Dalkeith came out in favor of Edinburgh, the issue was decided. As a result, in April of 1854, the Parliament granted £7,000 for the purchase of a site on Argyle Square to house the Industrial Museum (the collection of the Natural History Museum would continue to be displayed in the University until a permanent building could be completed). It had always been Wilson's goal to combine the two collections so that one could trace technological development from the natural world through finished products and one could see how the natural world served as a source for industrial development (Swinney 2016, 170–171). But since, in the beginning, the two collections were physically separated from one another, this would not be possible until the collections were united in a permanent building, which did not take place until after Wilson's death. It appears that Playfair had decided on Wilson to be the Director of the Museum by June of 1854 (Anderson 1992, 176). Edward Forbes, who was just moving to Edinburgh to become Regius Professor of Natural History expressed some interest in the position, but Playfair was against it. In any case, Forbes would die suddenly in November. Playfair and Wilson had been friends ever since they both were assistants in Thomas Graham's chemistry lab in London in 1838. As noted in an earlier chapter, David Brewster also argued that Wilson's work on color blindness, which was of great interest to railway and shipping companies that were installing colored lights as safety systems, played a role in his obtaining the Directorship of the Museum (J. Wilson 1866, 278).

Wilson's Vision

In discussing George Wilson's Industrial Museum, there is one big problem. Wilson died before the Museum was completed, and therefore, any discussion of the Museum must be a discussion of the Museum that existed in Wilson's head. But the Museum that existed in his head is still an interesting Museum. Geoffrey Swinney, who was a curator at the current National Museum of Scotland and who wrote a dissertation on the history of the Museum from its origins until 1939, has said that "even in the later part of the twentieth century the museum still carried the discernible imprint of Wilson's vision" (Swinney 2016, 187). An entry in the *Oxford Dictionary of National Biography* notes that the Museum "though changed in many ways from Wilson's conception, remains closer to the mid-Victorian pedagogic model than any other national museum in the United Kingdom" (Swinney 2016, 187). Therefore, it makes sense to analyze Wilson's vision and plan for his Museum. Throughout the remaining years of his life, Wilson dedicated a tremendous effort in establishing the Museum. His sister quotes him as saying: "I am determined," he sometimes said, "to let no day pass without doing something for my dear Museum" (J. Wilson 1866, 289). The clearest idea of his vision for the Museum can be found in his lectures and public talks he gave trying to publicize the Museum, raise money for the Museum, and obtain donations of objects for the Museum. He said on more than one occasion that the Museum served as a text, or textbook, for his lectures, so by analyzing his lectures, we can gain some insight into what he saw as the text of the Museum (G. Wilson 1855, 20).

In his *Inaugural Lecture* as Regius Professor of Technology, Wilson makes it clear that one of his major visions for the Museum is as a teaching and educational institution. As noted in the last chapter, he said: "With the Industrial Museum, this Chair stands in organic connection," but he went on to say: "The Museum, indeed, is intended to serve non-academic as well as academic Students, and as well as every stranger who may enter its doors," and would go on to say: "I must, so far as I can, and in all legitimate ways that I can, turn the Museum to the largest account as a means of education" (G. Wilson 1855, 19–20). He goes on to make clear that another major vision for the Museum is its role in economic and industrial development. His outline for an ambitious plan of ten sets of lectures included: "economic production and application of heat and light," which will include a discussion of natural and artificial fuels; a second division will be the "economic development of electricity," including its use in chemistry and the telegraph; a third division will focus on the "economic applications of light, as employed in the different modifications of Photogenic Art"; and a fourth division will be metallurgy, including the "economic applications of the metals and their alloys" (G. Wilson 1855, 21–22). He goes on to list other lectures on building materials, glass-blowing, pottery, textile manufacturing, paper-making, graphic arts, food production, animal products, and production of chemical products. This was an ambitious, or even overly ambitious, project, and he never organized it into single coherent set of lectures, but over the next four years, he was able to touch on most of these topics in his academic and in his public lectures.

The Industrial Museum and Agriculture

One of the first lectures Wilson gave concerning the Industrial Museum was to the Highland and Agricultural Society on January 16, 1856. It makes sense that he would present his ideas to them first since they were one of the earliest supporters of the idea of a national museum in Edinburgh, and once the Museum was established, they donated their rather large collection of minerals to the Museum. The title of his talk was "On the Relations of Technology to Agriculture" in which he lays out both the justification for the Museum and discusses how it will be arranged and organized and how it will benefit agriculture (G. Wilson 1857b). In justifying the need for the Museum at the beginning of this lecture, he links the Industrial Museum to a new interconnected world and says it is one of the institutions that have "become necessary by the altered condition of the world" (G. Wilson 1857b, 254). He goes on to argue that this altered condition is the result of new technology, especially the steamship, the railway, and the electric telegraph. One important consequence of this new interconnected world is that it has "revealed to us the unsuspected progress which the other great nations of the earth have been making in the industrial, as well as the other arts," but at the same time, those new technologies have "opened to us the whole civilised world, they have also let in the whole civilised world upon us" (G. Wilson 1857b, 254–255). He then goes on to argue that the interconnected world requires both learning and education. He says the "risk of being left behind ... led to the foundation of the Industrial Museum of Scotland, as enabling us to learn what we need to know, and to teach what we wish to be known" (G. Wilson 1857b, 255). In this sense, one of the Museum's roles will be to both gain new knowledge of technology from around the world and to disseminate the knowledge of technology that has been developed at home.

Wilson then goes on to discuss the organization of the Museum and how that organization is connected to his role as Professor of Technology. In his *Inaugural Lecture*, he spends time educating his audience to the new term "technology," and again defines it as "The 'Science of the Useful Arts,'" but also adds that technology contains not one, but many sciences (G. Wilson 1857b, 256). He lays out that it is the duty of the Director, "to collect, arrange, and preserve the objects, products and instruments of the industrial arts," and that the duty of the Professor "is systematically to expound on these within the walls of the University" (G. Wilson 1857b, 256). He notes that while the Museum is associated with the University, it is actually part of the Department of Science and Art under the Board of Trade and that a similar arrangement exists between the University and the Natural History Museum. This leads him to put forward an interesting idea. He says:

> When the new buildings are erected, it is intended to arrange the Natural History objects and the Industrial collections together, or in close proximity; so that, for example, the geological relations on the one hand, and the economical relations on the other, of coal, limestone, sandstone, ironstone, and the like, may be studied by those to whom the purely scientific or the purely industrial aspect of these minerals is alone interesting; whilst the many who desire to make themselves familiar with both aspects, will find the means of doing so under the same roof, and guided by a system of arrangement which contemplates their twofold study.
>
> (G. Wilson 1857b, 257)

The relationship between the Natural History Museum and the Industrial Museum had always been left somewhat vague, awaiting the completion of a permanent building, but it is clear that from the beginning it was Wilson's goal to integrate the two collections so as to be able to demonstrate his view of a cosmic and interrelated view of the world and his progressionist view of nature (Swinney 2016, 176). With the two collections integrated together, one could trace a modern industrial development back to pre-human geological history.

After laying out his organizational vision for the Museum, Wilson went on to explain how the Museum could have a direct impact on agriculture. He notes that the goal of agriculture is to produce on a given piece of land the largest amount of food in the shortest amount of time and to find ways to diminish the amount of land required and the amount of time required while increasing the amount of food produced (G. Wilson 1857b, 257). In order to do this, the agriculturist must have knowledge of a number of sciences, including meteorological studies of the seasons, winds, rain, and temperature. But since the basic laws of meteorology cannot be changed or manipulated, it would not be the role of the Industrial Museum to furnish illustrations of those natural laws, except to demonstrate ways to improve instruments, such as thermometers, barometers, and rain gauges. Rather:

> The great object, however, of the Industrial Museum will be to illustrate the application of science to the determination of the qualities of existing raw or initial materials, to the discovery or production of new ones, and to the derivation from familiar and from novel substances, of bodies serviceable in the arts.
>
> (G. Wilson 1857b, 258)

Interestingly, when he gave this talk, he had not formulated his paper "On the Physical Sciences which form the Basis of Technology," and seems to be defining technology as depending purely on the transformational sciences and does not appear to see a role for the observational and registrative sciences like meteorology.

Wilson then turns to a discussion of how technology can be of use to agriculture. He begins by raising the problem of simply building structures on a new farmstead. This requires decisions about building materials, such as the use of brick or stone in the construction of the buildings and further, what kind of bricks or stones. This will require some knowledge of say the properties of different materials such as hardness, durability, permeability to moisture, resistance to weather, and how quickly and cheaply they can be quarried and polished. At this point, Wilson discusses the fact that the Industrial Museum can be of benefit in making decisions about building materials because "one object of the Industrial Museum will be to collect specimens of all the building stones of Great Britain, and as far as shall prove possible, of the world" (G. Wilson 1857b, 259). But Wilson goes on to discuss a new role for the Museum beyond simply displaying building materials – the Museum will also contain an Analytical Laboratory. He states that:

> One object of the Analytical Laboratory attached to that Museum will be the analysis of such stones, with a view to discover how good and

bad building qualities stand related to the presence or absence of certain ingredients; and one important duty of the Professor of Technology will be to make the results of such researches known by lecture, exposition, and publication to the entire community.

(G. Wilson 1857b, 259)

This is an important new vision for the Industrial Museum. Not only will the Museum display objects and educate through lectures, but the Museum will also be a research laboratory and be actively involved in generating new knowledge based on its collections of objects.

Once the building stones are chosen, there becomes the issue of a choice of mortars. He notes there is an open question whether mortars made of sea-sand would always be wet so that dry walls would require pit sand. Again, the Museum will help to solve the problem through its collections and its laboratory. Concerning the issue of the type of sand used in mortar, Wilson says:

This last problem is still a vexed one, and it is only one among many, the solution of which will demand the examination of a large collection of old wall mortars, and many experiments on the properties of sand, lime, concrete, plaster, and cement.

(G. Wilson 1857b, 259)

Similar problems and solutions will occur in the choice of roofing materials, such as slates, tiles, metals, of thatch, and the Museum will play a role. He notes: "There of special and good qualities in each of them [roofing materials], and an Industrial Museum, by exhibiting systematically-arranged specimens of all, would furnish one important means of enabling agriculturists throughout the country to choose intelligently among them" (G. Wilson 1857b, 260).

After helping to solve problems of the building of an actual farmstead, Wilson argues that the Museum could help with other problems, such as choosing a water supply and deciding on what type of fuel will be used for heating, and again he sees the Museum providing an extensive collection of fuels arranged in terms of their locations, chemistry, efficiency, and costs. In addition, he sees the Museum as displaying grates, stoves, and furnaces along with a collection of means of lighting, including candles, oil lamps, gas lamps, and electric lights (G. Wilson 1857b, 261–262).

Besides aiding in the actual building of a farmstead, Wilson also saw an important role of the Museum in providing knowledge concerning the actual economics of running a farm. He notes that the production of wheat, hay, turnips, turkeys, and lambs require the use of other materials, and the cost of those other materials depends on the demand for those materials in other areas of the economy. Therefore, it is important for agriculturalists to have knowledge of manufactured products beyond agriculture. As an example, Wilson notes that animal bones have a high value in agriculture for producing fertilizers, but there are a wide variety of uses for animal bones, such as button-making, brush handles, glue-making, jelly-making, porcelain-making, drug-making, and refining sugar and metals. If in any of these other applications some substitute material is found, the price of bones will go down, so it is important for agriculturalists to be aware of developments in other markets and manufacturing processes so "the systematically-arranged bone products of a Technological Museum, would largely furnish him with the means of observing the development of rival professions" (G. Wilson 1857b, 264). Similarly, the new developments in the production and demand of certain chemicals, such as ammonia, nitrate of soda, and sulfuric acid, can have an important effect on the price of fertilizers. So, Wilson argued that, for example, in the case of sulfur:

> It would not then be a waste of time for an intelligent farmer to study in the Industrial Museum all the Technological relations of sulphur, for he has plainly an interest in its sources being multiplied, and its products or derivative being cheapened.
>
> (G. Wilson 1857b, 265)

Wilson summarized the role of the Industrial Museum in agriculture by saying:

> It cannot but be of service to a young man about to follow the profession of agriculture, to have the means of studying the sources, the objects, the processes, and the products of all the other manufactures of the country. It cannot but sharpen and strengthen the very faculties which will be most needed in his own calling. It cannot but expand his mind and quicken his sympathies, to possess an intelligent appreciation of the learning, patience, perseverance, faith, courage, ingenuity, inventive skills, and manipulative dexterity which have made other professions great and famous, and which, transferred to his own profession, will make it greater and more famous than it is.
>
> (G. Wilson 1857b, 265–266)

Again, this idea reflects Wilson's vision of an interconnected world.

The Objects of Technology and the Industrial Museum

In February of 1856 shortly after he gave his lecture on "The Relations of Technology to Agriculture," Wilson gave a set of two lectures to the Philosophical Institution of Edinburgh entitled "On the Objects of Technology and Industrial Museums." Both lectures centered on granite but included a range of other topics concerning technology and industrial museums. The first lecture focused more on the concept of technology and the second on industrial museums. As discussed in an earlier chapter, in this lecture, he begins by defining technology as "the science or doctrine of the peculiar practices of the arts," but then shortens the definition to "the science of useful art or the useful arts" (G. Wilson 1856a, 3). In his first lecture, a discussion of granite led to the various gems that had some relationship to granite. This in turn led to the idea that gems were only valuable as a substitute for money or for their use in brooches, pins, etc. For Wilson this raised the issue of utilitarianism and whether the utilitarian arts, like technology, were "stealing from us our imaginations" and "killing our conscience" (G. Wilson, 1856a, 5). He notes that if one follows his definition of technology, it could be claimed that it only has to do with how we obtain food and drink and how we are clothed and only fosters fine arts in terms of providing comforts or vanities, and that utilitarianism leads to the belief that the only true worth of an object is its monetary value. But Wilson goes on to argue that just such a focus on food, drink, and clothing is a positive value since the

> utilitarian does no more than declare that bread for the hungry, water for the thirsty, clothing for the naked, and homes for the houseless can be furnished to all, if men will but wisely use their faculties, and conquer that physical world which was given them to conquer.
>
> (G. Wilson 1856a, 6)

As Wilson sees it, satisfying basic human needs is no threat to the imagination and says "to feed, to clothe, to house the needy, are surely not acts which involve and invasion of their imaginations" (G. Wilson 1856a, 6). As Wilson sees technology, it has a moral role. Of industrialism he says, its "vocation far more is to relieve the wants of the poor than to minister to the luxuries of the rich; and we have the poor with us always" (G. Wilson 1856a, 7). This attitude would carry over to his vision of the Industrial Museum whose audience he saw as the working class not the upper class.

Wilson's second lecture focused more on the role of industrial museums. While the first lecture had focused on the physical and chemical properties of granite, in his second lecture, he focuses on how the ingredients that make up granite, such as quartz and feldspar, can be converted into useful products, such as glass, clay, and aluminum. He focuses most of the lecture on glass-making and discusses in detail the technical process involved in making glass objects of varying kinds. Before discussing how such glass objects would fit into an industrial museum, Wilson essentially puts forward a manifesto for the Industrial Museum. He says:

> An Industrial Museum is intended to be a repository for all the objects of useful art, including the materials with which each art deals, the finished products into which it converts them, drawings and diagrams explanatory of the processes through which it puts those materials, models or examples of the machinery with which it prepares and fashions them, and the tools which specially belong to it, as a particular craft. Such a

museum should also include illustrations of the progress of each industrial art from age to age; of its dependence on the sister arts, and the
extent to which it ministers to them; of its relation to the products of our
own country, and those of foreign lands; of the amount of wealth which
it consumes, circulates, and produces; of its healthfulness as a vocation
for the different sexes and ages; of its relation to good morals, and the
service which it can render the State by employing the needy, increasing
the comforts of the poor, advancing the civilisation of all classes, adding
to the material, intellectual, and more prosperity of the whole nation,
and through it, more or less the entire world.

(G. Wilson, 1856a, 14)

In this statement, Wilson highlights a number of important roles for the
Industrial Museum including demonstrating the progressionist nature of
technology, the interconnectedness of the industrial arts, the economic
impact of industrialization, and the moral role of technology in benefiting
the poor and advancing civilization.

As a concrete example of how the Industrial Museum would deal with the
various industrial arts, Wilson focuses on glass-making, and in doing so, we
see that important guiding principle of "Unity in Variety." Simply making
a glass object brings together a tremendous variety of sources of materials
and a tremendous variety of skills. He notes that the sand that goes into glass
may be brought from Scotland, the Isle of Wright, North or South America,
or Australia. The soda might originate in Great Britain or some of it might
come from Spain and Egypt, and even the soda that is locally produced is
done so through a process using sea salt that might be shipped from around
the world, and through the use of sulfur from Sicily. The potash for flint glass
might come from America, Canada, or Russia and the lead from Lanarkshire,
Dumfriesshire, and Cumberland. Finally, the various metals used to color the
glass, such as manganese, copper, tin, cobalt, silver, and gold, can again come
from dozens of different countries. Wilson then goes on to all of the skills
and machinery needed to build a glasswork. Masons, bricklayers, and potters
would be needed to build the structure and to provide the pots and kilns for
making glass. Machinery would be needed to grind and mix the materials
that go into the making of the glass. These machines would require the skills
of millwrights, carpenters, and engineers. Since many of the tools used in
glass-making are composed of iron, the skills of a blacksmith would also
be needed. In addition, chemists would be needed to analyze the chemicals
going into the glass and the glass itself, and artists would be needed to design
the glass objects. Finally, if the glasswork is large enough, it would require
managers and financiers. As an illustration of the fact that the wide variety
of materials from across the globe and wide variety of skills can all be interconnected in single glass object, he notes that if the Industrial Museum were
limited to only displaying glass, he "could include under that art all other
arts, because they are needful to it, as it is to them" (G. Wilson 1856a, 14).

It is not simply in its making that a glass object is interconnected to a wide number of geographical locations and to a wide number of other industrial arts, but a wide number of other scientific professions make use of the objects that glassmakers produce. For example, Wilson notes that astronomers depend on optical glass for its telescopes and botanists and anatomists need glass for their microscopes. Chemists are particularly dependent upon glass for the vessels they use to carry out their experiments. In fact, he says: "Chemistry may, in truth, so far as the greater part of it is concerned, be defined as the 'science of the glass vessel'" (G. Wilson 1856a, 14). But the natural philosophers are almost equally dependent on glass. Without glass, Boyle would not have had his air-pump, and without a glass prism, Newton would not have discovered the composition of light. Many of the instruments of natural philosophy like the thermometer, the barometer, and the original static electrical machines all required glass. Finally, photographers need glass for their lenses and plates, and botanists make use of glass to create glass houses, like those at Kew Gardens, to bring inside and study plants native to tropical climates. He goes on to note that non-scientific professions are dependent upon glass. Sailors need glass for their telescopes and sextants, and surveyors need glass for their theodolites, and watchmakers need glass for their magnifying glasses. All of these widely different professions are all linked to glass-making. As a final example of how there is some type of unity in the variety of glass objects, Wilson notes how the word glass has come to refer to a variety of different and distinctive objects: "The thirsty man calls his drinking vessel a glass. The sailor looks out for his landmarks with a glass. The beauty gazes into a glass. Best of all, the otherwise blind man ... calls his spectacles 'glasses'" (G. Wilson 1856a, 15).

Wilson ends his set of two lectures by introducing another role for the Industrial Museum – a role to display the work of women and to educate women in the industrial arts. He says in closing:

> I am especially anxious, in addressing this audience to enlist the sympathies of intelligent women in its [the Industrial Museum] behalf. They can contribute the graceful works of their own hands, which form some of the most highly prized objects in the English and Irish museums, and they can persuade those of the rougher sex, who come within the sphere of their influence, to give or procure objects for the industrial collection. I entreat them to do so, if from no other motive than from this, that they may thereby contribute to increase the means of giving an industrial education to women of the poorer classes, and to multiply the vocations which may keep them from starvation, misery, and crime.
>
> (G. Wilson 1856a, 16)

The Industrial Museum and Pharmacy

As part of his plan to publicize the idea of technology and the Industrial Museum, Wilson presented a short lecture to the Pharmaceutical Society in

March of 1856 with the title "On Pharmacy as a Branch of Technology." In the lecture, he reiterated many of his earlier ideas about technology, especially the interconnected nature of the industrial arts. He argued that "technology can best serve pharmacy by meeting it at the thousand points where it is dependent on the other arts, and on many sciences" (G. Wilson 1856b, 462). As examples of the varying arts and the varying skills that impact pharmacy, he notes that pharmacists need carpenters to build their cabinets and drawers, glassmakers to produce bottles and vials, potters to make pots, brass founders to make the scales, a chemist to produce the morphia and strychnine, distillers to make alcohol, botanists to provide the botanicals, and candlemakers to produce glycerin. But again, Wilson sees that besides this variety, there is also a unity. He says: "In short, every art is now dependent on every other art, and the object of technology is to insist on this, and to bring all arts together" (G. Wilson 1856b, 463). He concludes by referring to technology as a "republic of the arts," and asks the Society's help in donating objects to the Museum.

The Relation of Ornamental to Industrial Art

In order to reach a wider audience, Wilson gave a lecture on Christmas Eve, 1856 entitled "On the Relation of Ornamental to Industrial Art" that had been requested by the Art-Manufacture Association. In this lecture, Wilson provides a new role for the Industrial Museum. According to museum historian Carla Yanni: "The national museum would introduce Scots to the important concept that utility and beauty were allied, so they would learn not to fear beauty as a senseless luxury" (Yanni 1999, 98). In response to the Association's request, Wilson says: "I have gladly assented because my public vocation is to recommend the purely industrial arts, and I rejoice at the opportunity of illustrating how they wait upon the ornamental arts, and how these wait upon them" (G. Wilson 1857a, 3). The main problem that he wishes to address is whether utility and beauty must always be seen as opposed to one another, or if there is a way, they can be brought into some unity. Using the story of the "Beauty and the Beast," Wilson sees as a goal that

> whilst Beauty remains Beauty, the Beast becomes a graceful Prince, losing the clumsiness, but keeping the strength of his former state, and prince and princess join hands, each possessed of gifts which the other has not: Not like to like, but like in difference.
>
> (G. Wilson 1857a, 4)

He begins by admitting that the Scots are often seen as grim and rough people who have not cultivated the ornamental arts. He attributes this to the rejection of the sensuous in religious worship, the emphasis of the philosophical over the poetical, and the high value given to common sense.

According to Wilson, the Art-Manufacture Association's goal is to change the perception of the ornamental arts. He says:

> Its object is to promote the cultivation of Ornamental art, as distinguished, on the one hand from High art, and on the other from Industrial art; or what comes to the same thing, it seeks, by bringing down, so far as needs be, high art from its loftiest altitudes, and by raising, so far as needs be, industrial art from its lowest levels, to confer the grace of the one, and the strength of the other, upon ornamental art, the child of both.
>
> (G. Wilson 1857a, 8)

The key to this unification of the high arts and the industrial arts into the ornamental arts is what Wilson labels the art-workman. Without the art-workman, the industrial arts would never be "wedded" to the fine arts. The problem, as Wilson sees it, is that the industrial arts only care about utility and only accidentally produce beauty if there is no other way to make a useful object. If a glass vessel used by a chemist or oil merchant has a graceful shape, it is only because that was the easiest shape for the glass-blower to produce. On the other hand, the fine arts place little value in utility. While a stained-glass window will keep out the wind and rain, that is not the goal of the artist who created it. In Wilson's view, the fine arts represent an idealized reality symbolized by the imagination or the emotions. Because of these differences, the industrial arts stand apart from the fine arts, but for Wilson, this does not mean they are opposed to one another. The industrial arts do not prefer ugliness over beauty, and the fine arts are not hostile to utility. But if they are to be brought together in order to do so, there must be someone who can function as a middleman between the artist and the worker since it would not be realistic to imagine an artist and a worker functioning side by side (G. Wilson 1857a, 12–14). According to Wilson, this might be done by persuading an artist to become a worker or persuading a worker to become and artist. This would require someone attuned to the idealities of the fine art and the practicality of the industrial arts. Here Wilson reintroduces the idea of the eye and the hand that he introduced in his lecture "On the Physical Sciences that form the basis of Technology" that he presented as the opening lecture of the 1856–1857 academic year. He says that the art-workman would be someone "with an artist's eye, and an artificer's hand" (G. Wilson 1857a, 15). The goal of such an art-workman would be to beautify the common things. He sees a moral aspect to this and says:

> They must be beautified, because it is among them we live, and no occasional study of great works of art will sustain within us the sense of the beautiful, if at other times we are surrounded by mean and sordid things.
>
> (G. Wilson 1857a, 15)

Wilson seems to believe that one's environment is very important in shaping values. He gives the example of three young children being sent to Scotland, Ireland, and England for a period of time after which they would most likely acquire a Scottish, Irish, or English accent. Similarly, a person who is surrounded by "bright, harmonious, graceful things" will have much more appreciation for the fine arts than a person who is surrounded by "dull, ill-coloured, ungraceful objects" (G. Wilson 1857a, 18). Therefore, there is a great moral need for art-workmen who can beautify the common objects that surround people.

At the end of his lecture, Wilson challenges the idea that useful things cannot be beautiful and the more that you increase the beauty of an item the more you decrease its utility. As an example of the fact that beauty and utility can be brought together in a unitary form, he points to the example of the nautilus. He notes that the shape of the shell of the nautilus has been imitated by both artists and workers for its pure mathematical shape and the pearly luster of its shell. But while being a source of beauty, the nautilus "is a most perfect practical machine, at once a sailing-vessel and a diving-bell" (G. Wilson 1857a, 29–30). He continues by arguing that nature is the ultimate combination of beauty and utility. He says

> I need not further tell you that the flowers, the birds, the serpents, the stars, which, all through these galleries, are seen reappearing in works of art, are in their originals as much as the nautilus, combinations of utility, simplicity, and beauty. Not one of them is useless. Not one of them is ugly.
>
> (G. Wilson 1857a, 30)

Wilson ends his lecture again with another plea for a vision of unity. He argues that the science, art, literature, and religion can be seen to be interacting in a unity the same way the four chambers of the human heart function together (G. Wilson 1857a, 34). He concludes with a final image of unity. When all of the chambers of the heart are occupied with science, art, literature, and religion:

> Utility, Beauty, Philosophy, and Morality will be found, like the differently coloured rays of the sun, to produce by their union a beam of white light, in which any one of the four may be studied without harm to the rest, and Beauty can only help Morality.
>
> (G. Wilson 1857a, 340)

This may also connect to Bacon's idea of the *lumen siccum* (the dry light) is the best soul (Bacon 1937, 178).

The Industrial Museum and Commercial Enterprise

Wilson's final public lecture concerning the Industrial Museum was given on December 4, 1857, at the request of the Company of Merchants of the

City of Edinburgh and entitled, "The Industrial Museum of Scotland in its Relation to Commercial Enterprise." Clearly from the title this lecture was aimed at making an economic role for the Museum by connecting it to commerce. Early in his lecture he says the Museum was to be a "tower of refuge in commercial storms," and a "castle stored with ammunition and weapons of commercial warfare," and it was not to be a "holiday institution" (G. Wilson 1858, 9). As he argued in his lecture to the Highland and Agricultural Society, he sees the need for the Industrial Museum as a arising from new conditions in the world brought about by the new transportation and communication technologies which as a result: "The great globe has seemed before our eyes to contract into smaller dimensions," and the "entire globe is now an open market-place and bazaar for every nation" (G. Wilson 1858, 11–13; Swinney 2016, 184). In such a new globalized world and at time of India's First War of Independence and financial crises in America, Wilson believed that the Museum "will largely help us to hold recovered India, and to diminish the recurrence of American panics," and that it "will increase our civilisation and add to our power to civilise the rest of the world" (Swinney 2013, chap. 7, n. 6). This "civilizing" rhetoric was closely tied to Great Britain's concept of empire. This civilizing rhetoric can also be seen in Wilson's view of the development of technology and how the objects in the museum should be displayed. According to Marinell Ash, Wilson, like others of his time, associated human progress with technological development. As we have seen earlier, Wilson was attracted to a progressional view of history and human development where parts of the world with less developed technology represented earlier stages of both human and technological development which places like Great Britain had already passed through (Swinney 2016, 183). Given this view, Wilson argued that the objects in the Museum needed to be organized historically. He said:

> that the conditions of those Arts must be illustrated historically, at least as practices in our own and other Civilised Countries within recent periods; and that their contemporary condition in less Civilised Nations must also be illustrated (examples from Africa, India, Chine, &c).
>
> (Swinney 2016, 183)

Planning the Industrial Museum and Building its Collections

As a physical structure Wilson saw the Museum composed of four elements: "1. An ample Exhibitional Gallery; 2. Laboratory and Workshop; 3. A Library; and 4. Systematic Lectures" (J. Wilson 1866, 314). As he saw it the Museum itself would hold collections of raw materials, intermediate stages of production, and finished products along with the tools, apparatuses, and machines that produced them. In doing so, it would embody "ideas and technological solutions important to the intellectual and industrial progress of Scotland," and in addition, "the development and communication of technological know-how was a vehicle for social improvement and advantage" (G. Wilson 1858, 39–40). Museum historian Carla Yanni argues that Wilson saw the Museum as imitating the Museum of Practical Geology in London, but along with displays of economic geology, it would also include economic botany and economic zoology reflecting Wilson's belief that world commerce dealt "very largely with mineral, vegetable, and animal substances" (Yanni 1999, 98). This idea was reflected in Wilson new expanded emblem for the Museum. In this lecture in addition to the eye in the middle of an open hand, Wilson added

> a circle, to imply that the museum represents the industry of the whole world; within the circle, an equilateral triangle, the respective sides of which shall denote the mineral, vegetable, and animal kingdoms, from which industrial art gathers its materials ....
>
> (Swinney 2016, 166)

Aside from discussing the purpose and goals for the Industrial Museum, Wilson also talked about building the Museum collection. In a letter to his sister Jean telling her of his appointment, he said:

> On the objects of my Museum, and the Arts connected with them, my plan will be as follows: – If a Shoemaker comes to the Museum, I'll talk to him about nothing but Hats, and screw information out of him about Shoes. When a Hat-maker arrives, I will pour into his ears all the learning I have acquired from the Shoemaker, and extract from the Hatter information to give the Cobbler on his next visit.
>
> (J. Wilson 1866, 279)

Building a collection proved to be a challenge especially since much of the knowledge associated with producing industrial objects was tacit knowledge and not written down in books, but the strategy he outlined above appeared to overcome some of those challenges. In his lecture on "The Industrial Museum of Scotland in its Relation to Commercial Enterprise," he discussed how he used his strategy to obtain objects for his collection along with the knowledge of how they were produced. He said:

> one of the chief, and I confess unexpected, obstacles which I encounter in seeking to fill the Industrial Museum with examples of Art, is the humble estimate which men form of their own callings. I cannot persuade a shoemaker that shoes are of interest to any but shoemakers and the barefooted public, although he looks with eager curiosity at my collection of hats in all their stages. I tried in vain to induce a very intelligent glassmaker to send me certain specimens of glass, till I showed him

a full series of illustrations of brush-making. His eyes brightened with interest, and he admired the ingenious and unexpected devices which an art strange to him revealed. Well, said I, be sure the brush-maker will be as much interested in your glass as you are in his brushes, so send me what I ask. I cannot, accordingly help inferring that a stranger's curiosity will often make up for his defective experience, and that the Industrial Museum would secure his services for all the arts it represented.

(Swinney 2016, 181)

Building the Museum's collection benefited from the donations of already existing collections from various societies (Anderson 1992, 179). As already noted, during the Museum's first year, the Highland and Agricultural Society donated its collection of minerals as well as its collection of agricultural equipment and machinery. That year the Museum also received a collection of building stones from the Architectural Institute of Scotland and machine models from the Royal Society for the Arts. Later the Museum would also receive duplicate specimens from the Museum of Practical Geology in London and received a number of objects from Great Exhibition of 1851. Wilson was particularly concerned to obtain objects and sets of objects that could show an entire manufacturing process. R.G.W. Anderson notes that Wilson was able to get a number of donations from important companies, such as pottery from Wedgwood and India rubber from Charles Mackintosh and Company (Anderson 1992, 179). Wilson's Museum also benefited from the growing British Empire. Many Scottish nationals were involved in overseas ventures, either as military officers, missionaries, East India Company officials, or people in private business, and they were recruited to obtain articles for the Museum. An important source of objects for the Museum was George Wilson's brother Daniel, who in 1853 had become Professor of History and Literature at University College, Toronto. As one of North America's leading anthropologists, Daniel had many connections across Canada and was able to secure a number of ethnographic articles and helped George to obtain a large collection of sub-Arctic Canadian artifacts from the Hudson's Bay Company (Swinney 2016, 187). In addition, the many of the students in his classes taught at the university came from a wide variety of backgrounds, including "general manufacturer, architect, engineer, farmer, merchant, baker, tanner, sugar planter, sugar refiner, teacher, doctor, and clergyman" who could also serve the Museum (Swinney 2016, 185). In the Museum's Annual Report for 1856, Wilson wrote:

A correspondence accordingly has been opened with agents in different quarters of the globe, which is likely to prove fruitful. A small collection of the objects manufactured by the Bassouta Africans has already been acquired. The simple forms of spinning machinery still in use in Madeira are on the way to this country. From Western Africa, Chili and the Red Indian districts (especially Minnesota) of the United States, intelligent men interested in the Museum have engaged to send examples of the native manufacturers of those countries.

(Swinney 2016, 184)

In the four-year period before his death in 1859, Wilson managed to acquire 10,350 objects (J. Wilson 1866, 290). Anderson has categorized the collection and found that there were 512 acquisitions with one donation from Godfrey Wedgwood consisting of 601 items (Anderson 1992, 180). Most of the collection fell under the categories of Industrial Arts, Relics and Antiquities, and Geology, with small numbers in the categories of Foreign Ethnology, Natural History, and Science and Engineering. His friend Lyon Playfair was able to help him acquire a collection of historic chemical apparatus from the University. But the collection had a number of odd items, as if they were from a cabinet of curiosities. Swinney notes that some of the first items acquired included:

timepieces and other metal objects damaged by fire, Indian minerals, a section of jointed submarine cable, a cast of a fossil footprint, a series of lead pipes from Edinburgh tenements illustrating the process of decay and damage by rats, a shepherd's crook from the Cheviot Hills, and a mollusk shell discovered in the mortar in the wall of a crypt of Glasgow Cathedral.

(Swinney 2013, chap. 5, n. 6)

Wilson not only had to acquire a collection for the Industrial Museum but also had to serve as curator of that collection. Swinney, who himself was a curator at the National Museum of Scotland, discussed how Wilson had to establish his own curatorial style (Swinney 2016, 174). As we have seen, it was intended that the Industrial Museum would be combined in some way with the Natural History Museum, but they would each approach the purpose of collecting in different ways. In an 1855 article in *The Scotsman*, that Swinney attributes to Wilson, an example is given of the difference between an industrial museum and a natural history museum when it came to dealing with a dead horse:

The Museum of Natural History would include either a horse in life, or its skin stuffed, as its type in death, but the Industrial Museum looks chiefly with interest to the carcass rejected with horror by the keeper of the natural history collection. Its ears and hoofs appear a Prussian blue, with numerous applications, the iron shoes as swords and guns, the bones as gelatin and isinglass, the skin as leather, and the blood as manure, the flesh as food for domestic animals.

(Swinney 2016, 174)

As another example of contrast, Swinney notes that both the Natural History Museum and the Industrial Museum were interested in collecting electric eels and skates, but the Natural History Museum saw them as examples of the natural fauna of West Africa, while Wilson saw them as examples of the first electric machines that could be used for electrotherapy (Swinney 2013, chap. 5, n. 8).

Based on his view that technology should focus on principles that guide the useful arts and not on the practice of those arts, Wilson's goal for Museum collections was not to collect single objects but to collect a range of objects that would demonstrate the principles that guided the manufacturing process (G. Wilson 1855, 4; Swinney 2016, 174–175). This led him to value common, even imperfect, items over rare objects since the imperfections would help to demonstrate the manufacturing process. Swinney quotes from a report by Wilson where he says of a Wedgwood ceramic piece with a speckled glaze: "Mr. Wedgwood supposes this fault is owing to the copper not being sufficiently mixed with the glaze at some particular point in the fusion" (Swinney 2016, 175). In another letter concerning a chipped teapot that the Museum received from the South Kensington Museum, Wilson says: "the fracture demonstrates all the more distinctly the absence of vitreous glaze which characterise this lustrous ware, and makes it none the less acceptable in the Industrial Museum" (Swinney 2016, 175–176).

During the summer of 1859, Wilson had a new idea concerning ordering the collection of the Industrial Museum. As we have seen, George Wilson had a very close relationship with his older brother Daniel throughout his life. Even as children they had been involved in collecting botanical and geological specimens (J. Wilson 1866, 9). Daniel had attended the University of Edinburgh and by the 1840s had established himself as one of the first scientific archeologists in Scotland, or even the entire English-speaking world (Trigger 1992, 55–75). In the 1840s, Daniel became involved in a project of the Society of Antiquaries of Scotland to transform its collection into a National Archeological Museum for Scotland. Marinell Ash has argued that Daniel used a system of "ordered groupings analogous to chemical elements" to organize the new museum (Swinney 2016, 187). Given the close familial and intellectual connection between the two brothers, it is not surprising that in 1859, George suggested incorporating the archeological collection under the same roof as the industrial and natural history collections. In a letter to Daniel, who in 1853 had become Professor of History and Literature at University College in Toronto and would go on to become President of the University of Toronto, Wilson wrote: "Oh, that they would put the antiquaries' museum under the same roof as mine and make you Professor of Archeology and let us devise monograms and plan museums and lecture rooms, etc., as in the old schoolboy days" (Swinney 2016, 188). The plan did eventually take place but not until 1985 when the National Museum of Scotland was established.

Wilson's Final Year

Tragically George Wilson would not live to see his Industrial Museum built. In 1858, Parliament had granted funds, and some preliminary surveying was done in the summer of 1859 but George Wilson died on November 22, 1859. As we have seen, Wilson suffered a number of health issues, especially after the partial amputation of his foot. His most serious problem was with tuberculosis. We have already seen that Wilson was at first reluctant to accept the directorship of the Industrial Museum because he was suffering another health crisis at the time. At the time of his appointment as Regius Chair of Technology, he wrote to a close friend: "Had Her Majesty consulted my doctors, she would have given me a sofa rather than a chair" (J. Wilson 1866, 285). Throughout his entire tenure as Director of the Industrial Museum, he suffered a series of health crises, mostly associated with his tuberculosis, but he always seemed to be able to recover and, at least in his letters, always seemed to maintain a positive attitude. In September of 1859, he had enough strength to attend the BAAS meeting in Aberdeen which included Prince Albert, and Wilson was able to join in one of the "Red Lions" dinners and met with Michael Faraday and William Thompson (J. Wilson 1866, 347–349). In November, he was well enough to begin his lectures for the 1859 session, but on November 17th, he traveled to Glasgow to deal with a patent-infringement case for a friend. The weather was cold and rainy which led to him developing a cough. His lungs were always weak from the tuberculosis, and within a short time, his cough had developed into pleurisy, bronchitis, and pneumonia, and on November 22, 1859, George Wilson died. On Monday November 28th, there was a large public funeral attended by students and faculty of the University, the Lord Provost and Magistrates, members of the Royal Scottish Society of Arts, the Pharmaceutical Society, the Chamber of Commerce, the Merchant Company, and the Philosophical Institution (J. Wilson 1866, 371–373). His burial took place at the Old Calton burial ground, and his grave was marked by a large antique cross that contained at its base the emblem of the circle, triangle, hand, and eye that he had proposed for the Industrial Museum.

References

Anderson, Robert G.W. 1992. "What is Technology? Education through Museums in the Mid-Nineteenth-Century." *British Journal for the History of Science* 25: 169–184.

Bacon, Francis. 1937. "Advancement of Learning." In Francis Bacon, *Essays, Advancement of Learning, New Atlantis, and Other Pieces*, edited by Richard Foster Jones, 171–238. New York: Odyssey Press.

Forbes, Edward. 1853. *On the Educational Uses of Museums*. London: HMSO.

Foucault, Michel. 1973. *On the Order of Things: An Archaeology of the Human Sciences*. New York: Vintage Books.

Kriegel, Lara. 2007. *Grand Designs: Labor, Empire, and the Museum in Victorian Culture*. Durham, NC: Duke University Press.

Swinney, Geoffrey N. 2013. "Towards an Historical Geography of a 'Natural' History Museum: The Industrial Museum of Scotland, the Edinburgh Museum of

Science and Art, and the Royal Scottish Museum, 1854–1939." PhD diss., University of Edinburgh.

Swinney, Geoffrey N. 2016. "George Wilson's Map of Technology: Giving Shape to the 'Industrial Arts' in Mid-Nineteenth-Century Edinburgh." *Journal of Scottish Historical Studies* 36: 165–190.

Trigger, Bruce G. 1992. "Daniel Wilson and the Scottish Enlightenment." *Proceedings of the Society of Antiquaries of Scotland* 122: 55–75.

Wilson, George. 1855. *What is Technology?* Edinburgh: Sutherland and Knox.

Wilson, George. 1856a. *On the Objects of Technology and Industrial Museums: Two Lectures.* Edinburgh: Sutherland and Knox.

Wilson, George. 1856b. "On Pharmacy as a Branch of Technology." *Pharmaceutical Journal and Transactions* 15: 457–463.

Wilson, George. 1857a. *On the Relation of Ornamental to Industrial Art.* Edinburgh: Edmonston and Douglas.

Wilson, George. 1857b. "On the Relations of Technology to Agriculture." *Transactions of the Highland and Agricultural Society of Scotland*, new series: 254–268.

Wilson, George. 1858. *The Industrial Museum of Scotland in Relation to Commercial Enterprise.* Edinburgh: R. and R. Clark.

Wilson, George and Archibald Geikie. 1861. *Memoir of Edward Forbes, F.R.S.* London: Macmillan and Co.

Wilson, Jessie Aitken. 1866. *Memoir of George Wilson.* London: Macmillan and Co.

Yanni, Carla. 1999. *Nature's Museums: Victorian Science and the Architecture of Display.* Baltimore, MD: Johns Hopkins University Press.

7 Epilogue

Victorian Science and Technology after George Wilson

George Wilson died at a time when Victorian science and technology were beginning to undergo significant changes. Two events would prove to be particularly influential. Just two days after Wilson's death on November 22, 1859, Charles Darwin published his *On the Origin of Species*, and a little over two years after Wilson's death, Prince Albert died on December 14, 1861. As we will discuss in more detail, Darwin's theory would provide a new model for science, what has been called scientific naturalism, that would replace the theistic naturalism championed by Wilson and other early Victorians (Stanley 2015; Dawson and Lightman 2014; Turner 1993; Desmond 1989). At the same time, the death of Prince Albert would remove one of Britain's leading advocates for scientific and technical educational reform as well as an advocate for British technology and industrialization. Between 1843 and his death in 1861, Prince Albert served as the President of the Royal Society for Arts, Manufacturers, and Commerce which was the driving force behind the Great Exhibition of 1851, and between 1846 and his death, he served as the Chancellor of the University of Cambridge. These, and a number of other causes, led to the development of British science and technology in the second half of the nineteenth century to differ from the early Victorian period. In that early period, Great Britain was a leader in technological development as a result of the Industrial Revolution. As we have seen, the railway and the telegraph would dramatically change the world. But in the second half of the century, the major new technological developments, such as the telephone, electric lighting systems, the internal combustion engine, and automobiles, all originated in other countries, such as the United States, Germany, and France. Germany, in particular, with its new developments in chemical and steel, was challenging Great Britain to become the world's leading industrial power. While Darwin's theory of evolution would redefine biology, many of the new areas of research in biology, such as bacteriology, were developed by Louis Pasteur in France and Robert Koch in Germany. Great Britain still played an important role in the development of physics during the second half of the nineteenth century because of the work of Michael Faraday and James Clerk Maxwell in electromagnetism and the work of Lord Kelvin and W.J.M. Rankine in the science of energy, but even in energy science, new

DOI: 10.4324/9781003212218-7

contributions were being made by Germans, such as Rudolf Clausius, Hermann von Helmholtz, and Ludwig Boltzmann. Lord Kelvin in his famous "Two Clouds" speech given in 1900 at the Royal Institution could still claim that British physicists, particularly Isaac Newton through his three laws of mechanical motion and James Clerk Maxwell through his equations of electromagnetism, had provided a system to explain much of the physical world. But his "two clouds" (problems explaining the Michaelson-Morley experiment and problems explaining black-body radiation) would lead to German scientists developing the theory of relativity and quantum theory which would undermine both Newton's and Maxwell's theories.

Museums

Given the changes that would take place in the second half of the nineteenth century, it is probably not surprising that George Wilson's early Victorian vision of science and technology would not continue to be completely realized in the later Victorian period. But a number of the developments that took place still reflected parts of his vision. As noted in the last chapter, Wilson would not live to see the actual construction of the Industrial Museum of Scotland. The Museum was constructed in a series of phases. The first phase was opened in 1866 followed by a second phase opened in 1875 and a third phase opened in 1889 (Swinney 2016, 186). Even when the first phase opened, the name had been changed to the Edinburgh Museum of Science and Art and its focus had changed from that originally envisioned by Wilson. The Department of Science and Art in London appointed Thomas Archer, who had helped establish a Trade Museum in Liverpool, as "Superintendent," rather than "Director" of the Museum, and unlike Wilson, he had no formal connection with the University of Edinburgh and was allowed to give only one introductory lecture each year (Swinney 2016, 186; Anderson 1992, 182). In 1858, Lyon Playfair, who had championed the Industrial Museum and George Wilson, left the Department of Science and Art to take up the Chair of Chemistry at the University of Edinburgh. As a result, Henry Cole became in charge of the development of the Museum and began to change its focus. As we have seen, the original purpose of the Industrial Museum was to stimulate British manufacturing through improving production. This would be done through lectures, laboratories, and the collection of materials that demonstrated the process of turning raw materials into finished manufactured items. This required not just the acquisition of a single item, but an entire array of items illustrating the entire production process. This resulted in the acquisition of a large number of specimens that were difficult to store until a permanent building could be constructed, and even then, it might quickly outgrow that building (Swinney 2016, 184–186).

With Playfair gone from the Department of Science and Art, Henry Cole, influenced by the success of the Great Exhibition of 1851, began to refocus his museums, including South Kensington and the Edinburgh Museum away from a study of production and onto the idea of design (Anderson 1992, 183). That is, rather than trying to stimulate British manufacturing by improving manufacturing, Cole's new idea was to stimulate manufacturing by increasing consumption which would require educating the British public in the decorative arts. Instead of collecting series of pieces, representing the manufacturing process, museums would now collect outstanding pieces of design that would raise the tastes of the British public. This new role for museums led to a physical change in Wilson's original design for the Museum. Carla Yanni has shown that as the Museum developed, it came to be dominated by a central "Great Hall" with an arcade around the upper balcony (Yanni 1999, 102–106). The building became more focused on exhibition and display than on research. Yanni references museum theorist Tony Bennett's idea that museums began to resemble Department Stores and incorporated some of the same architectural elements, where items for consumption were put on display, but the visitors themselves also were put on display through the use of open arcades and balconies (Yanni 1999, 106–108). While Wilson's vision of the role of the Industrial Museum of Scotland was not realized, as noted in the last chapter, it was said that the Museum would carry his imprint into the later part of the twentieth century and would remain closer to the mid-Victorian pedagogical model than any other British museum (Swinney 2016, 187).

Although Wilson's vision of the role of a museum as a laboratory for the improvement of technology was not realized in Edinburgh, some of his ideas came to be reflected in other museums. Dale Idiens in a study of ethnology collections argued that George Wilson, his older brother Daniel, and Lt.-Gen. Augustus Henry Lane-Fox Pitt-Rivers were part of a new movement in how museums dealt with the material culture of non-western civilizations (Idiens 1994, 5). He notes that instead of focusing on collecting and displaying individual rare objects, the Wilson brothers and Pitt-Rivers collected a range of typical objects and arranged them in sequences. The Pitt-Rivers Museum, established at Oxford University in 1884, would become the museum most similar to Wilson's idea of the Industrial Museum of Scotland, although there is no evidence that Pitt-Rivers was influenced by Wilson. During his time in the Grenadier Guards, Pitt-Rivers became interested in weapons after being assigned to test the new pattern 51 (Miné) rifle that was replacing the smooth bore muskets (Thompson 1977, chap. I). By 1852, he became interested in collecting firearms and soon he extended his collection to other weapons, focusing on primitive weapons. This, in turn, led him to include a wide range of ethnological objects, including tools, jars, and musical instruments, but weapons were always at the center of his collecting. Influenced by the new ideas of Charles Darwin, Pitt-Rivers came to see that he could use his collection to study the biological roots of weapons and to uncover the principles that governed their development (Thompson 1977, 35; Basalla 1988, 17–21). As we have seen, Wilson argued for the biological roots of technology, but there is no firm evidence that Pitt-River was following Wilson rather he seems to have been directly influenced by Darwin's ideas through the Eth-nological Society of London (Thompson 1977, 31–33). In a series of lectures on "Primitive Warfare," delivered at the Royal United Services Institution between 1867 and 1869, Pitt-Rivers outlined his Darwinian theory on the development of warfare and weaponry (Pitt-Rivers 1906, 45–186). His lectures were divided between the Stone Age, the Bronze Age, and the Iron Age, which is similar to the three-age system devised by the Danish arche-ologist Christian Thomsen in the early nineteenth century and which was followed by the Museum of the Society of Antiquaries of Scotland and served as the organizational structure for Daniel Wilson's book *The Archeology and Prehistoric Annals of Scotland*, but again no firm connection to Wilson can be established (Trigger 1992, 61–62). Like George Wilson, Pitt-Rivers saw biological roots to human weapons. Early armor imitated animal hides and scales, spears imitated horns, tusks, and antlers, and serrated knives imitated sharks' teeth (Pitt-Rivers 1906, 57–68). But Pitt-Rivers went beyond the role of imitation to argue that the development of weapons followed a Darwinian model. Pitt-Rivers argued:

> In the earliest stages of art, men would of necessity be led to the adop-tion of such varieties by the constantly differing forms of the materials in which they worked. The uncertain fractures of flint, the various curves of trees out of which they constructed their clubs, … would lead them imperceptibly towards the adoption of fresh tools. Occasionally some form would be hit upon, which in the hands of its employer, would be found more convenient for use, and which be giving the possessor of it some advantage over his neighbours, would commend itself to gen-eral adoption. Thus a process resembling what Mr. Darwin in his late work has termed "unconscious selection," rather than be premeditation or design, men would be led on to improvement.
>
> (Pitt-Rivers 1906, 96)

If these variations produced better weapons, they would be adopted (Pitt-Rivers 1906, 95–96). This theory led Pitt-Rivers to argue that museums should arrange weapons so as to demonstrate how small variations led to new types of weapons. His most famous example was to show how an Australian aboriginal wooden sword that was sometimes thrown at an enemy could, through a series of variations brought about by following the natural grain of the wood, eventually lead to the boomerang and that variations in construct-ing boomerangs could lead to a Malaga (a type of pick), or a waddy (a type of hatchet) (Pitt-Rivers 1906, 121–126). Given his theory concerning the evo-lution of weapons, he argued that the collections in the Pitt-Rivers Museum differed from most other museums. Most museums collected and displayed objects because of their value, uniqueness, or beauty, but he chose to collect and display objects that were typical rather than rare (Pitt-Rivers 1906, chap. 2).

For Pitt-Rivers, the displays of the linear sequences of the development of weapons fulfilled his idea of the purpose of his Museum which was to

> trace out … the sequence of ideas by which we find mankind has advanced from the condition of the lower animals to that in which we find him at the present time, and by this means to provide really reliable materials for a philosophy of progress.
>
> (Pitt-Rivers 1906, 9–10)

It should be noted that Wilson expressed some similar ideas when he argued that an industrial museum must show the "historical illustrations of the progress of the useful arts" (Swinney 2016, 183). Given his belief that new and improved inventions seem to be the result of variations on existing inventions, Pitt-Rivers seemed to be arguing that museums could function as laboratories for the improvement of technology since such improvements could not arise from studying a single artifact but require studying a large number of artifacts from different cultures and different historical periods (Pitt-Rivers 1906, 91). This seems very similar to Swinney's argument that Wilson saw the Industrial Museum as "a technology for teaching technology" (Swinney 2016, 177). We have also seen that a central element of Wilson's idea of his Museum would be a laboratory to study and improve technological development. Modern critics would certainly reject many of both Wilson's and Pitt-Rivers ideas about the organization of museums, especially the linear theory of technological progress and the racist theories of non-modern and non-western cultures, but both Wilson and Pitt-Rivers raise interesting points about the role of museums as laboratories and not just centers of display, consumption, and entertainment. Science laboratories were not unknown, especially in chemistry, but the idea of an engineering laboratory was somewhat rare. The idea of museums as engineering laboratories might seem to foreshadow the important change that took place in technological development with the introduction of industrial research laboratories in the late nineteenth and early twentieth centuries.

Engineering Education

Wilson's vision for the role of technology in the University of Edinburgh ended with his death and was not realized again until several years later and then not in the same form that he had imagined. As we have seen, the introduction of technology or engineering into British universities did not always go smoothly. Unlike the French and Germans, British engineers were suspicious of theory and thought the best way to teach engineering was through the apprenticeship system (Jenkyn 1870, 198–204). As we have seen, the two oldest chairs of engineering (or technology) founded in Scotland were not at the request of the universities' faculty but were established by the Crown and in both Glasgow and Edinburgh were met with jealousy from the professors in the natural sciences who saw such chairs as taking away their students. Wilson's course on technology was quite popular, attracting more than eighty students, which was very large compared with similar courses at other British institutions (Birse 1983, 68). As such, very shortly after Wilson's death, the Senate of the University established a committee composed of the Professors of Natural History, Chemistry, and Natural Philosophy who recommended that a replacement for Wilson not be appointed but that the instruction he was giving be divided among the other professors (Anderson 1992, 182). The year after Wilson's death Sir David Brewster, Vice-Chancellor and Principal of the University, made a strong argument in his inaugural address for the need of civil engineers to be trained in the sciences having been influenced by the establishment in Paris of the *École centrale des arts et manufactures* (Birse 1983, 68–71). Even with Brewster's support, it would not be until 1868 that he was able to find funding for a Chair of Engineering and soon after Fleeming Jenkin, who was Professor of Engineering at University College London and who would go on to teach the great writer, Robert Louis Stevenson while at Edinburgh, was appointed Professor of Engineering (Birse 1983, 94–110).

In one sense, this was a realization of one of Wilson's visions of science and technology. As we have seen, Wilson's earliest definition of technology was: "Science in its application to the Useful Arts" (J. Wilson 1866, 279). As we have also seen, later Wilson would outline the range of science, upon which technology is based (G. Wilson 1857). This idea that technology, or engineering, was based on science was not widely supported by many British engineers. Even Jenkin, in his inaugural address to the University of Edinburgh, cautioned against requiring engineers to be taught as much theory as was taught in France (Jenkyn 1870, 203). Therefore, the fact that by the twentieth-century engineering education was firmly based on science and taught through university courses is a fulfillment of one of Wilson's visions of science and technology. But Wilson's vision was not fulfilled, for the most part, in the form that he originally imagined.

As we have seen, Wilson saw all of the sciences playing a role as the basis for technology, but he saw a special role for biology. As noted earlier, the set of lectures he designed in technology was organized more like a course in natural history than in natural philosophy, in that the three years were divided into mineral technology, vegetable technology, and animal technology (J. Wilson 1866, 288). Except for a brief discussion of the telegraph, there was little discussion of the physical sciences, such as mechanics, or thermodynamics and there appears to be almost no role in mathematics (Birse 1983, 67–68). In contrast, the recommended courses under Jenkin included two years of mathematics, two years of natural philosophy, chemistry, mechanical drawing, and surveying, along with engineering courses that included statics, dynamics, hydrodynamics, kinematics of machinery, the steam engine, and strength of materials, among many other topics (Birse 1983, 100–101). These are topics that would be treated in any engineering school today, but it would be difficult to find anything similar to Wilson's topics on "The Plant as a manufacturing agent," or "Bones, Horns, Shells, Corals, their Mechanical and Chemical applications" (Birse 1983, 67–68). The topics in Jenkin's syllabus were drawn almost exclusively from the physical sciences, but other engineering courses in Great Britain did include topics such as geology (King's College, University College London, Royal School of Mines, Trinity College Dublin, Royal College of Science for Ireland, Owens College, Manchester), mineralogy (King's College, Royal School of Mines, Trinity College, Dublin, Owens College, Manchester), and paleontology (Royal School of Mines, Royal College of Science for Ireland) (Institution of Civil Engineers, 1870, 3–16).

Technology and Biology

Wilson's idea that technology should be based on biology was also not an aberration during the nineteenth century (Channell 1991, chap. 5). He was the most vocal advocate for interpreting technological development in organic or biological terms, but he was not alone in this approach. In 1842, before Wilson was calling attention to the connection between biology and technology, Thomas Ewbank, who would go on to become the U.S. Commissioner of Patents from 1849 to 1852, noted in his book on hydraulic machines that: "Few classes of men are more interested in studying natural history, and particularly the structure, habits and movements of animals, than mechanics; and none can reap a richer reward for the time and labor expended upon it" (Ewbank 1842, 258). He saw many connections between the current technology of his day and biology. He notes:

> The flexible water-mains ... by which Watt conveyed fresh water under the river Clyde were suggested by the mechanism of a *lobster's tail* – the process of tunneling which Brunel has formed a passage under the Thames occurred to him by witnessing the operations of the *Teredo*, a testaceous *worm* covered with a cylindrical shell, which eats its way through the hardest wood –and Smeaton, in seeking the form most adapted to impart stability of the light-house on the Eddystone rocks, imitated the contour of the bole of a *tree*.
>
> (Ewbank 1842, 258)

In a book entitled *The World a Workshop*, published in 1855, the same year Wilson was appointed to his chair, Ewbank makes many of the same points that Wilson would make in his subsequent lectures and publications, although there is no evidence that Wilson knew of Ewbank's work. For example, Ewbank begins by saying that chemistry, physics, botany, and zoology "will pour, and continue to pour forth, new elements, combinations, forms, forces, and motions" (Ewbank 1855, vii). Like Wilson, Ewbank divides the world into the three Linnaean categories of mineral, vegetable, and animal, what he calls three store-houses of matter and gives particular attention to the role of organic life. He notes: "The factory would have been a failure if its operations had been confined to mineral stock, hence the Vegetable store-house," referring to the important role of wood in construction, and went on to note that the animal department: "Furnishes matter of different forms and properties to those drawn from the vegetable and mineral stores," referring to the use of animal products for human fabrics, such as wool feathers and silk (Ewbank 1855, xiii). Also, like Wilson, he argues that nature provides man hints and "applications in profusion, to guide him in producing creations of his own" (Ewbank 1855, 152). Finally, like Wilson, Ewbank sees: "The Divine Character Displayed in Natural Mechanisms," and that "it is the first and last duty of man ... to imitate him [the Creator]" (Ewbank 1855, 179).

Both Wilson and Ewbank made the connection between technology and biology very explicit, but other engineers made the use of biology in their work more implicit and used biology to develop technologies that on their surface were quite mechanical. This was particularly true of a number of engineers who were developing a theory of mechanisms that governed the transmission and modification of motion within machines (Ferguson 1962; Reuleaux 1876). One of the most important biological ideas to play a role in the development of mechanical technologies was the concept of biological classification. During the nineteenth century, the study of mechanisms was transformed by the application of various more natural systems of classification. In earlier periods, it was common to consider every machine as a separate and distinct whole, consisting of parts peculiar to that machine (Reuleaux 1876, 9). That is, machines were classified according to the purpose for which each machine was designed with little emphasis placed on the underlying relationships between different types of machines (Willis 1870, viii). Even when engineers began to distinguish single mechanisms from specific machines, they were usually studied for their own sake, and only incidentally for their more general application to other machines.

By the end of the eighteenth century and the beginning of the nineteenth century, the search began for a more systematic way to classify machines. An important step was taken by Gaspard Monge at the newly established École Polytechnique when he argued that a study should be undertaken of the elements of a machine that converted one type of motion into another since the most complicated machines were simply combinations of these elements (Willis 1870, viii). Between 1806 and 1808, his followers, Jean Nicolas Pierre Hâchette, Phillipe Lanz, and Agustin de Betancourt, developed a system of classification that was based on the conversion of one type of motion into another. Using two characteristics, one describing the type of motion, such as rectilinear, circular, or curvilinear, and another characteristic describing the direction of motion, such as continuous or alternative mechanisms, could be classified into twenty-one possible categories (e.g., continuous circular motion changed to alternating circular motion). This classification system closely paralleled similar classification systems used in natural history, such as Carl Linneaus' binomial system.

In his book, *The Order of Things*, Michel Foucault argues that: "Throughout the eighteenth century, classifiers had been establishing character by comparing visible structures, that is, by correlating elements that were homogeneous" (Foucault 1973, 226). But, in the nineteenth century, there was a movement from what Foucault calls "described structure" to "classifying character," and this "transformation of structure into character, was based upon a principle alien to the domain of the visible – an internal principle .... This principle .... is *organic structure*" (Foucault 1973, 227). If we follow Foucault, the classification system introduced by Monge and his followers in the early nineteenth century was a move toward a more organic theory of mechanisms since it depended upon the function of the internal mechanisms and not on the outward design of the machine.

By the middle of the nineteenth century, there were attempts at establishing even more "naturalistic" classification systems based on the functional aspect of mechanisms. For example, in 1841, Robert Willis, Jacksonian Professor at Cambridge University, focused his attention on the given relationship of motions created by a mechanism rather than changes in speed and direction of a given motion. For Willis, a clock-work mechanism was classified as a device that maintained the angular velocity of the hands in a ratio of twelve to one and kept their direction of rotation similar. The operation of the mechanism was now seen as independent of the given motion since the clock-word mechanism would still function if the motion were back-and-forth rather than continuous (Willis 1870, xiv). Willis's classification system would fit Foucault's idea of an organic system of classification since it linked superficial characteristics of a machine, such as modes of connection (e.g., rolling contacts, sliding contacts, gears), to the essential function of the mechanism, such as maintaining velocity or directional ratios.

By the second half of the nineteenth century, theories of mechanisms were being discussed and debated in terms of biological models. One example is Franz Reuleaux's theory of mechanisms. He was Professor of Engineering at Zurich and later Director of the Royal Technical School of Charlottenburg. Reuleaux criticized earlier classification systems as not being sufficiently natural. He said:

> Monge's classification, however natural it appears, does not in the first place correspond to the real nature of the matter. Did it so – did it resemble, for instance, the classifications of Linnaeus and Cuvier in organic nature – it would like them be able to make its footing firm.
>
> (Reuleaux 1876, 17)

Reuleaux's focus was away from individual mechanisms and toward the relationships that existed between certain types of mechanisms. This would fit into Foucault's argument that:

> the link between one organic structure and another can no longer, in fact, be the identity of one or several elements, but must be the identity

of the relation between the elements (a relationship in which visibility no longer plays a role) and the functions they preform; ....

(Foucault 1973, 218)

Reuleaux's approach to mechanisms treated mechanisms as part of a continuous process of relationships. For example, rather than treating a crank joined to a rocker arm by a connecting rod, as a simple combination of three parts, he argued that the frame, bedplate, or even the floor, which held the axle or the crank and the rocker arm in their positions, formed a fourth member of the combination, and the whole system could be seen as a closed chain or circuit (Suplee 1905, 816). When viewed as a circuit, any element member might be considered as fixed the others as moving. As a result, one could describe mechanisms that were "inversions" of other mechanisms, but all of these "inversions" would result in mechanisms that had some "natural" relationship between them.

An underlying assumption of the attempts to develop more naturalistic classification systems for mechanisms was the idea that such systems could not only help understand existing mechanisms, but also they could lead to the discovery of new mechanisms similar to the emergence of new species from existing species. Reuleaux argued that a theory of mechanisms would not "deserve the name of Science" until it was able to develop "the true classification of its own material" (Reuleaux 1876, 20). He went on to claim that when that happened, they would "furnish the means for arriving at new mechanisms..." (Reuleaux 1876, 20). Although Reuleaux did not specifically mention the term evolution, a number of other engineers did begin to think of technological development in evolutionary terms (Basalla 1988). Herbert Spencer, who was trained as an engineer before he turned to philosophy, believed that the entire universe should be governed by evolutionary principles. He believed that all processes in the universe could be seen as the evolution from the simple to the complex, or from the homogeneous to the heterogeneous through a process he called integration. In his book *First Principles* (1864), he argued that technology, like the rest of nature, should be governed by evolution and said: "The progress from small and simple tools to complex and larger machines, is a progress in integration" (Spencer 1900, 297). As an example of moving from the simple to the complex, he uses the example of how simple Archimedean machines could evolve into complex textile machines. He said:

> Among what are classed as mechanical powers, the advance from the lever to the wheel-and-axle is an advance from a simple agent to an agent made up of several simple ones. On comparing the wheel-and-axle, or any of the mechanical appliances used in early times with those used now, we see that in each of our machines several of the primitive machines are united. A modern apparatus for spinning or weaving, for making stockings or lace, contains not simply a lever, an inclined plane,

a screw, a wheel-and-axle, joined together, but several of each – all made into a whole.

(Spencer 1900, 297)

Spencer saw this process of integration continuing beyond individual machines. He says: "A much more extensive integration is seen in every factory. Here numerous complicated machines are all connected by driving shafts with the same steam-engine – all united with it into one vast apparatus" (Spencer 1900, 298).

One of the most significant examples of the use of a biological theory of technology during the second half of the nineteenth century was the work of Karl Marx. Although known primarily for his economic and political theories, Marx was also a historian and philosopher of technology, and his ideas about technology played an important role in his economic and political theories, especially in his major work, *Das Kapital*. In 1841, Marx wrote a Ph.D. dissertation on the "Difference Between the Democritean and Epicurean Philosophy of Nature" and he continued to have an interest in the philosophy of nature throughout his life. His interest in the biological world is also reflected in his interest in Darwin's *On the Origin of Species*. Most importantly, in preparation for writing *Das Kapital*, Marx attended a course on technology taught by Professor Robert Willis, who we have seen was playing an important role in developing a naturalistic classification system of mechanisms (Marx 1978, 362).

Marx saw technology closely connected to biology. In his chapter in *Das Kapital* on "Machinery and Modern Industry," he noted:

A critical history of technology would show how little any of the inventions of the 18th century are the work of a single individual. Hitherto there is no such book. Darwin has interested us in the history of Nature's Technology, *i.e.*, in the formation of the organs of plants and animals, which organs serve as instruments of production for sustaining life. Does not the history of the productive organs of man, organs that are the material basis of all social organization deserved equal attention?

(Marx 1977, 372)

Marx saw *Das Kapital* as just such a study of "the productive organs of man." He draws a parallel between a study of the history of human technology and biological history. He says the "relics of bygone instruments of labor possess the same importance for the investigation of extinct economic forms of society as do fossil bones for the determination of extinct species of animals" (Marx 1977, 179–180). He clearly saw the "instruments of labor," in biological terms. He argued that the instruments "of a mechanical nature, taken as a whole, we may call the bones and muscles of production," and such instruments as pipes, tubs, and jars that store the materials for production "we may in a general way, call the vascular system of production" (Marx 1977, 180).

Marx's critique of the capitalistic system rested on the idea that modern technology has turned craftsmen into simple workers by taking away their tools and incorporating those tools into modern machines. For Marx, the reason that machines could take over the work of skilled craftsmen was that machines were, in some sense, alive. He said:

> An organised system of machines, to which motion if communicated by the transmitting mechanism from a central automaton, is the most developed form of production by machinery. Here we have, in the place of the isolated machine, a mechanical monster whose body fills whole factories, and whose demon powers, at first veiled under the slow and measured motions of his giant limbs, at length breaks out into the fast and furious whirl of his countless working organs.
>
> (Marx 1977, 381–382)

The factory was, for Marx, the epitome of modern technological production and the greatest threat to the workers. The main reason for this threat was that the factory and technology had taken on an organic character and had a life of its own. In one of his descriptions of the modern factory, Marx quotes Andrew Ure, an early nineteenth-century Scottish physician, geologist, chemist, and industrial theorist and the author of *The Philosophy of Manufacturers* (1835), who called the factory "a vast automaton, composed of various mechanical and intellectual organs, acting in uninterrupted concert for the production of a common object, all of them being subordinate to a self-regulating moving force" (Marx 1977, 418–419). Ironically Ure saw this in positive terms since if freed children from doing heavy labor and it gave other labors free time to read books. But Marx saw the factory in much darker terms. In his version of the factory, "the workmen are merely conscious organs, [and] coordinate with of unconscious organs of the automaton," and they do not have leisure time, rather "the attendants [of the machines] are reckoned more or less all 'Feeders' who supply the machines with materials to be worked" (Marx 1977, 419–420). Marx was not using biological terminology simply as a metaphor for the machine, for Marx machines incorporated actual vital or living force and the fact that they were truly alive made them a danger to the working class.

On the other side of the Atlantic, biological models were also influencing the development of technology. A leading advocate of the importance of biology to technology was Robert Henry Thurston, who was one of America's most distinguished engineers and engineering educators (Durand 1929). Thurston helped to establish the engineering curriculum, first at Stevens Institute of Technology and later at Cornell University. He wrote several classic textbooks on engineering and helped to organize the American Society of Mechanical Engineers and served as its first president. In a series of papers, presidential addresses, and a remarkable book entitled *The Animal as a Machine and a Prime Motor* (1894), he argued for the importance of biological ideas for future improvements in technology.

Thurston argued that the greatest challenge facing civilization was "making the utilization of the forces of nature, more general, more efficient, and more fruitful" (Thurston 1894, 84). He said:

> Could the engineer ... find a way of producing steam–power at a fraction its present cost; could he transform heat energy directly and without waste into dynamic; could he find a method of evolution of light without that loss now inevitable in the form of accompanying heat; could he directly produce electricity, without other and lost energy ... the advancement ... of the race would be inconceivably altered.
>
> (Thurston 1894, 84–86)

But according to Thurston, technology had "apparently reached a point beyond which he [the engineer] can see but little opportunity for further improvement, except by slow and toilsome and continually limited progress" (Thurston 1894, 85). Thurston argued that the "only recourse" to going beyond the limits of our traditional mechanically based technology would be "scientific research and the study of nature's own methods" (Thurston 1894, 86). He said: "Nature accomplishes many of the tasks that man is about attempting, and has been holding up to him the solution of his problems throughout the ages" (Thurston 1894, 86). More specifically, for Thurston, the solution to a new technology was to be found in the biological world. He noted that in a living body

> mechanical power is exerted at less cost in potential energy supplied than in the steam-engine; heat is evolved as the product of combustion or other action at a low temperature; light is produced by glow-worms and fire-flies without sensible loss in accompanying thermal energy; and electricity is produced with similar wonderful economy.
>
> (Thurston 1894, 45)

Although Thurston was at times too optimistic – for example, he thought living systems could evade the second law of thermodynamics – he believed that nature's processes were much more economical than the technologies developed by humans, and in order to improve our existing technologies, the engineer "has but to imitate her" (Thurston 1894, 90–91). Thurston concludes his book saying:

> It seems more than probable that it is to the mysteries and lessons of life that the chemist, the physicist, the engineer, must turn in seeking the key that shall unlock the still unrevealed treasures of the coming centuries. These constitute nature's challenge to the engineer.
>
> (Thurston 1894, 97)

As we have already seen, Pitt-Rivers argued that military technology had its roots in the animal world with armor imitating hides and scales, clubs imitating hooves, and knives imitating teeth. We have also seen that he went on to argue that improvements in weapons followed something similar to Darwin's theory of natural selection which governed the biological world. There is no evidence that Eubank, Willis, Reuleaux, Pitt-Rivers, Marx, or Thurston were directly influenced by the work of Wilson, but their work indicates that Wilson's ideas about the relationship between technology and biology were not unique but reflected an important alternative view of technology during the Victorian period.

Unity in Variety: From Theistic Science to Scientific Naturalism

Finally, Wilson's vision of the unity and uniformity of science continued to be realized in the second half of the nineteenth century but in a significantly different context than the one used by him to formulate his vision of "Unity in Variety." As we have seen throughout this book, Wilson constantly focused on a unity that existed behind the great variety of the natural world. In this, he was not alone. We have seen that Alexander von Humboldt focused his *Cosmos* on the idea of "Unity in Diversity," and Mary Somerville focused her *On the Connexion of the Physical Sciences* on the ability to "unite detached branches [of science] by general principles," and John Herschel's *Preliminary Discourse* claims that the only useful facts are those "which happen uniformly." We also saw that the transcendental naturalists focused their work on the concept of a "unity of plan," and that James Clerk Maxwell's work was guided by the idea of the "unity of the universe." Finally, we saw that many of the *Bridgewater Treatises* drew on the concept of a unity of nature. This concept of the unity of nature had a correlation to the idea of the uniformity of nature – that at different times and places, similar causes would produce similar effects.

We have also seen that during the first half of the nineteenth century, this focus on unity was closely tied to religion and led to the development of natural theology or more broadly a theology of nature or a theistic science. Using the writings of James Clerk Maxwell, Matthew Stanley argues that believers in theistic science saw unity and uniformity as a crucial element of science and saw the source of that unity and uniformity as God (Stanley 2015, 35–52). It was through the unity of natural laws that God communicated his existence to human beings (Stanley 2015, 42). While some theologians argued that imposing the concepts of unity and uniformity onto God was limiting God's power to intervene in the world, the majority of supporters of theistic science held to the belief that unity and uniformity only made sense in a world governed by God, since without some imposed order, the world would be one of chaos (Stanley 2015, 36, 52).

While the concept of unity that was at the center of Wilson's vision of science would serve as a fundamental principle of science, by the second half of the nineteenth century, the context for the belief in that idea of unity had undergone a significant change. In 1892, Thomas Henry Huxley popularized the term "scientific naturalism," as a new principle of science in the second half of the nineteenth century (Dawson and Lightman 2014, 1). While Huxley claimed credit for coining the term, historians have discovered that it was being used in America in the 1840s by evangelicals as a disparaging term for science that was opposed to scriptural doctrines (Dawson and Lightman 2014, 4–5). By the middle of the century, the term came to signify an opposition to supernaturalism and spiritualism, which seems to have been one of Huxley's meanings, but he also often used the term scientific naturalist to refer to an expert practitioner as opposed to an amateur scientist. But even here, there was an underlying critique of religion since many of the amateur scientists in Victorian Britain were clergymen. While the term was used in a variety of ways during the second half of the nineteenth century, its most common usage was as a reference for a science that was no longer based on religious beliefs. Frank Turner has argued that the scientific naturalists saw themselves as "fighting a major battle against the adherents of traditional natural theology" (Turner 1993, 117).

For many scholars, this war between a theistic science and scientific naturalism was brought about by the publications of Charles Darwin, beginning in 1859 (just days after Wilson's death) with *On the Origin of Species* but continuing with *The Variation of Animals and Plants under Domestication* (1868), *The Descent of Man* (1871), and *The Expression of the Emotions in Man and Animals* (1872) (Turner 1993, 118–123). In all of these works, Darwin was providing naturalistic explanations for ideas that had been a fundamental part of natural theology, especially Paley's "argument from design." But recent scholarship, especially the work of Peter Bowler, has shown that Darwin's work did not bring an end to the ideas of natural theology (Bowler 2009, chaps. 6, 7). Darwin's theory of natural selection left open a number of issues. At the center of natural selection was the fact that there are "chance" variations in the offspring of organisms, and if those variations confer some benefit in terms of survival or attracting a mate, the competition for either food or a mate will select out those beneficial variations which will then be inherited by the next generation. The problem then, and even today, is the meaning and source of those chance variations. Chance can have two meanings. One is that chance simply refers to human ignorance. That is, there is some pattern of purposeful designs in the variations but we, as humans, are simply ignorant of those designs. Second, chance can simply mean that there is no design in nature and, therefore, no designer. Darwin was not clear what he meant by chance designs and in fact seemed to change his mind by the end of his life when he seemed open to the idea that there might be some pattern behind the variation. Further confusing the issue was that his leading advocate in Britain, Thomas Henry Huxley, interpreted chance as a lack of design, while his leading supporter in America, Asa Gray, assumed that there was a pattern behind the variation and, therefore, some type of design. Darwin did seem to be critical of natural theology and saw his theory of natural selection as providing an alternative to Paley's argument from design and in *The Descent of Man*, he seemed to be providing naturalistic explanations of what were thought to be unique human abilities and attributes that were used by natural theologians as evidence of God's design (Turner 1993, 119–120).

As a number of scholars have argued, the roots of scientific naturalism are broader than just Darwinian evolution, and included utilitarianism, positivism, socialism, phrenology, and the rise of professionalism (Dawson and Lightman 2014, 10–16). Robert M. Young has also argued that while scientific naturalism seemed to be opposed to Christianity, it actually had many secular elements that paralleled elements of Christian theodicy. According to Paul White: "James Moore defined 'scientific naturalism' as a rival worldview or theodicy for a group of 'dissident intellectuals' who opposed the authority of the established Church" (White 2014, 237). Other scholars have pointed to the social origins of scientific naturalism through such organizations as the X Club, a dining club founded in 1864 that included Huxley, John Tyndall, and Herbert Spencer, among others, which lent support to reforming the Royal Society, reforming education, especially scientific education, support of Darwinism, but it also had ties with liberal Anglicanism (Stanley 2015, 30–32). Matthew Stanley goes so far as to suggest that the X Club played an important role in the ability of scientific naturalism to eventually win over theistic science by gaining control of the scientific establishment (Stanley 2015, 30). Huxley's idea of scientific naturalism was also spread through his membership in the Metaphysical Society, a social group suggested by James Knowles, editor of the *Contemporary Review* in 1868 (White 2014, 220–222). Huxley was a founding member and coined the term "agnostic" at its first meeting. The founding members were mostly clergymen whose initial plan was to fight against science and skepticism, but it quickly turned into something like a broad debating society with a diverse membership of both clergy and secularists. Paul White has shown how within the Metaphysical Society the boundaries between scientific naturalism and more theistic approaches to science began to become blurred arguing that Huxley's original definition of "agnosticism" was seen as a way to bridge differences between scientific naturalists and more religious groups (White 2014, 237–238).

The idea of a bridge between theistic science and scientific naturalism provides a new way to view how Wilson's vision of science became realized in the second half of the nineteenth century. If the idea of agnosticism could be seen as a bridge, an even more important bridge was the central role of the unity and uniformity of nature in both theistic science and scientific naturalism. We have already seen that unity and uniformity played an important role in theistic science, but Stanley has argued that it played an equally important role in scientific naturalism (Stanley 2014, 242–262; Stanley 2015, chap. 2). While the theists saw the unity and uniformity of nature as both a theological and a scientific concept, the scientific naturalists saw uniformity as fundamental to a secular view of science since it would prevent God from intervening into nature through such things as miracles (Stanley 2014, 248–253).

Unity and uniformity raised problems for both theists and scientific natural-ists. Theists did have to explain the "miracle problem" and did so by arguing that miracles might not be the violation of some natural law but simply the result of some as yet unknown law of nature, such as seeing a balloon as defying gravity if the law of buoyancy was not known. Or a miracle might follow some scientific law and not be the result of some supernatural force, but the miracle was in the timing or intent of what was ultimately a natural event (Stanley 2014, 253). Scientific naturalists, like Huxley, made almost the same argument to explain what were often seen as nonuniform events, like earthquakes and volcanic eruptions (Stanley 2014 245). Such events could be likened to a clock that was designed to explode gunpowder at a preset time. Ironically, evolution posed as many problems for the scientific naturalists as for the theists. While evolution provided a naturalistic explanation for Paley's argument from design, it also raised questions about the uniformity of nature since species seemed to be changing. In his early days, Huxley, who would go on to become "Darwin's bulldog," opposed evolution, since the best evidence for fixed laws would be fixed species (Stanley 2015, 57). As a way around the problem, a number of scientists developed what has come to be called theistic evolution. For example, in Britain, the Duke of Argyll argued that creating a new species through natural laws was still divine creation and as we have seen, Asa Gray in America held that there was an unseen pattern and direction behind the chance variation that drove natural selection (Stanley 2015, 60).

Given the fact that both theists and scientific naturalists both saw uni-formity as a central concept in their view of science, Stanley has argued that it served to bring the two groups together and smoothed the transition from theism to scientific naturalism (Stanley 2014, 253–254). He argues that while there was a significant shift in what was thought to be the roots of science and the source of scientific laws, the actual *practice* of science for each group remained relatively unchanged. As he notes, no one attacked theistic scien-tists, such as James Clerk Maxwell, or scientific naturalists, like John Tyndall, by trying to get them to lose their jobs or threatening to stop publishing their works.

Eventually, the scientific naturalists won the debate over the nature of science but not because one version of science was true and the other false. Stanley suggests that the scientific naturalists came to prevail over the theists because they were able to control the education of the next generation of sci-entists through such organizations as the X Club (Stanley 2015, chap. 7; Stan-ley 2014, 254–258). Members of the X Club, and especially Huxley, found themselves in positions of power when it came to the new educational reforms taking place in the 1870s and 1880s which gave them influence over scientific education through publishing textbooks, giving public lectures, recruiting students to study science in the universities, and in training elementary teach-ers of science. Most importantly, because the new educational reforms placed a large influence on an exam system rather than actual teaching, many of the scientific naturalists played a large role in writing and grading science exams which came to reflect the secular views of scientific naturalism. While scientific naturalism would come to define science in the late nineteenth century and would provide a model of science very different from Wilson's, his vision of "unity in variety" would continue to live on after him.

References

Anderson, Robert G.W. 1992. "What is Technology? Education through Museums in the Mid-Nineteenth-Century." *British Journal for the History of Science* 25: 169–184.

Basalla, George. 1988. *The Evolution of Technology*. Cambridge: Cambridge University Press.

Birse, Ronald. 1983. *Engineering at Edinburgh University: A Short History, 1673–1983.* Edinburgh: University of Edinburgh.

Bowler, Peter. 2009. *Evolution: The History of an Idea*, 25th Anniversary Edition. Berkeley: University of California Press.

Channell, David F. 1991. *The Vital Machine: A Study of Technology and Organic Life.* New York: Oxford University Press.

Dawson, Gowan and Bernard Lightman. 2014. "Introduction." In *Victorian Scientific Naturalism: Community, Identity, Continuity,* edited by Gowan Dawson and Bernard Lightman, 1–24. Chicago: University of Chicago Press.

Desmond, Adrian. 1989. *The Politics of Evolution: Morphology, Medicine, and Reform in Radical London.* Chicago: University of Chicago Press.

Durand, William F. 1929. *Robert Henry Thurston.* New York: American Society of Mechanical Engineers.

Ewbank, Thomas. 1842. *A Descriptive Historical Account of Hydraulic and Other Machines of Raising Water.* London: Tilt and Bogue.

Ewbank, Thomas. 1855. *The World a Workshop, or the Physical Relationship of Man to the Earth.* New York: D. Appleton and Co.

Ferguson, Eugene. 1962. "Kinematics of Mechanisms from the Time of Watt." *United States National Museum Bulletin* no. 228. Washington, DC: Smithsonian Institution.

Foucault, Michel. 1973. *The Order of Things: An Archaeology of the Human Sciences.* New York: Vintage Books.

Idiens, Dale. 1994. "A Wider World." In *A Wider World: Collections of Foreign Ethnography in Scotland,* edited by Elizabeth Kwasnick, 3–7. Edinburgh: National Museums of Scotland.

Institution of Civil Engineers. 1870. *The Education and Status of Civil Engineers in the United Kingdom and Foreign Countries.* London: Institution of Civil Engineers.

Jenkyn (*sic*), Fleeming. 1870. "On Engineering Education." In *The Education and Status of Civil Engineers in the United Kingdom and Foreign Countries,* 198–204. London: Institution of Civil Engineers.

Marx, Karl. 1977. *Capital,* vol. 1, edited by Friederick Engels, translated by Samuel Moore and Edward Aveling. New York: International Publishers.

Marx, Karl. 1978. "Letter from Marx to Engels, January 28, 1863." *In The Essential Marx: The Non-Economic Writings,* edited and translated by Saul Padover, 361–364. New York: New American Library.

Pitt-Rivers, Lt.-Gen. 1906. "August Henry Lane-Fox.*The Evolution of Culture and Other Essays*, edited by John .L. Myres. Oxford: Clarendon Press.

Reuleaux, Franz. 1876. The *Kinematics of Machinery: Outlines of a Theory of Machines*, translated by Alex B.W. Kennedy. London: Macmillan and Co.

Spencer, Herbert. 1900. *First Principles*, 6th ed. New York: D. Appleton Co.

Stanley, Matthew. 2014. "Where Naturalism and Theism Met: The Uniformity of Nature." In *Victorian Scientific Naturalism: Community, Identity, Continuity*, edited by Gowan Dawson and Bernard Lightman, 242–262. Chicago: University of Chicago Press.

Stanley, Matthew. 2015. *Huxley's Church and Maxwell's Demon: From Theistic Science to Scientific Naturalism*. Chicago: University of Chicago Press.

Suplee, Henry .H. 1905. "Franz Reuleaux, In Memorium." *Transactions of the American Society of Mechanical Engineers* 26: 816.

Swinney, Geoffrey N. 2016. "George Wilson's Map of Technology: Giving Shape to the 'Industrial Arts' in Mid-Nineteenth-Century Edinburgh." *Journal of Scottish Historical Studies* 36: 165–190.

Thompson, M.W. 1977. *General Pitt-Rivers: Evolution of Archeology in the Nineteenth Century*. Bradford-on-Avon: Moonraker Press.

Thurston, Robert H. 1894. *The Animal as a Machine and a Prime Mover, and the Laws of Energetics*. New York: John Wiley and Sons.

Trigger, Bruce G. 1992. "Daniel Wilson and the Scottish Enlightenment." *Proceedings of the Antiquaries of Scotland* 122:55–75.

Turner, Frank M. 1993. *Contesting Cultural Authority: Essays in Victorian Cultural Life*. Cambridge: Cambridge University Press.

White, Paul. 2014. "The Conduct of Belief: Agnosticism, the Metaphysical Society, and the Foundation of Intellectual Communities." In *Victorian Scientific Naturalism: Community, Identity, Continuity*, edited by Gowan Dawson and Bernard Lightman, 220–240. Chicago: University of Chicago Press.

Willis, Robert. 1870. *Principles of Mechanism: Designed for Students in Universities, and for Engineering Students Generally*, 2nd ed. London: Longmans.

Wilson, George. 1857. "On the Physical Sciences which Form the Basis of Technology." *The Edinburgh New Philosophical Journal*, newseries 5: 64–101.

Wilson, Jessie Aitken. 1866. *Memoir of George Wilson*, new ed. London: Macmillan and Co.

Yanni, Carla. 1999. *Nature's Museums: Victorian Science and the Architecture of Display*. Baltimore, MD: Johns Hopkins University Press.

Index